Amplifiers

Peter Rodmell G3ZRS

Peter

73

Peter G3ZRS

FEATURING

History and evolution of amplifiers and valves

How to make HF VHF UHF amplifiers

How to repair amplifiers in detail

Aerial tuning units

First published 2014

ISBN: 978-0-9930351-0-4

Page Layout by
Highlight Type Bureau Ltd, Bradford BD8 7HB

Printed and bound in Great Britain by
The Amadeus Press, West 26 Business Park
Cleckheaton, BD19 4TQ

CONTENTS

Chapter 1

Introduction

Danger warning

Brief description of author

Amateur products made by Linear Amp UK

Commercial amplifiers made by Linear Amp UK

Introduction

I have written this book with help from my good friend Val Gardner and my wife Gwen M0GWA who helped me to compose, edit and prepare for publishing. I have self published this book and funded it by choice so I do apologise for any mistakes or errors you find and I will be pleased to know about any.

It is intended to help with the understanding of amplifiers and help those who intend to build or repair HF, VHF and UHF amplifiers. The choice of valves is huge and is still the most economical way of generating a stronger signal from your radio station.

An amplifier is the powerhouse of your radio station. Many keen DXers who have large aerials arrays will tell you , go down to 100 watts you can call for ever, but add even 400 - 500 watts and you will add 7dB to your signal which makes all the difference between being heard or not.

I have always been fascinated by steam engines. In a lot of ways I look on a valve amplifier in a similar fashion. You give it fuel, it gets hot and glows red and then it gives you the power you require to get to where you are going, either by the radio waves or on the rail track!

If you drive it too hard or give it a bad load it complains and gets hot and sometimes even overheats!

A Ham Station is not complete without the Powerhouse. Life is too short for QRP!

The choice of valves is now vast and while most people think the valve is a thing of the past, the majority of amateur radio stations still use valve amplifiers.

Peter Rodmell G3ZRS
Founder Linear Amp UK

DANGER WARNING

RF Amplifiers run at very high voltages and temperatures, touching the wrong parts can kill as well as giving you a very nasty RF burn or heat burn.

For your own safety, do not work on amplifiers live and do not override the internal interlocks unless you are 100% confident of what you are doing.

I have worked on High Power Radio Frequency amplifiers for the last 40 years, repairing many as well as finishing them for final tuning, always remembering safety first.

Be careful and do enjoy experimenting.

I must express how important it is to work safely.

Happy building and repairing.

Peter Rodmell G3ZRS
Founder Linear Amp UK

A brief bit about the author Peter Rodmell G3ZRS

I was born on the 1st September 1950 and acquired my ham radio licence in 1970 after several years of short wave listening.

I became interested in ham radio in the boy scouts, my Scout Leader Mike Ellis G3PJR was a great enthusiast, we erected aerials on the top of the local Church and operated the radio 'Jamboree on the Air' weekend around 1964. We contacted other Scout groups all over the world and from that day I was hooked on Amateur Radio. At the time, my nearest radio amateurs were John Dunnington G3LZQ and Vic Lindgren G4BYG, both only lived a short distance away and were a great help in my early days Ham radio. Whilst at school I bought an old Eddystone 358X and later a National HRO with ham band coils, the bees knees! I was then a keen SWL from 1964.

During my electrical apprenticeship (described later), I was taught every form of industrial Electrical Engineering. During this time I was also very fortunate to learn other skills including joinery, welding, turning and blacksmith skills, thanks to my employer BTDB British Transport Docks Board, now known as ABP, all of which were very useful skills for making amplifiers, aerials and masts!

The RSGB and my opinion: - G3ZRS

The RSGB is the basis of the amateur radio community in the UK. I joined the RSGB as a SWL in the 1960s and have always felt that they are a bond for all radio amateurs in the UK and some abroad. For youngsters there has to be a way forward and with the workers and volunteers through the RSGB there is a good foundation to keep what we have and improve amateur radio facilities.

We, as radio amateurs, must remember that many other services are trying to take parts of our spectrum especially at VHF, UHF and microwaves. The RSGB through their national and international meetings with world organisations help keep what we have and improve.

We in the UK compared with most other countries in Europe have been one of the first to get many of our new bands which include:-

First to get 6 metres in Europe and get high power for the band

Some other countries are now getting 4m which we have had for years

WARC implanted quickly with the RA

Expansion of the 40m band to 7.2 MHz

60m / 5MHz

The VLF new frequencies

If I think the RSGB could do anything more for us it would be to get rid of the 400 watt anomaly. Most HF Dxers have amplifiers which will give more than 400 watts. Around the world 1 kW has become a norm and in the USA and Canada 1500 watts. When amateurs take part in a contest or are chasing that rare, weak DX station you want to be on a level playing field. I am sure it is not a problem with the RA. When most radio hams say 400 watt, they really mean 1 kW or more. We have always, in this country, had to be conservative with all we do but maybe sensibility could now be used. If you live on a farm or in the country you would not do any harm by using 1 kW.

If your aerial is very close to other domestic equipment, it would be stupid to use high power. The responsibility is initially with you to try to solve any interference problems that you might have. It is almost the same as when it is safe to drive fast on a main road! Given the choice, you would not speed through a housing estate or pass a school but it is safe on a good motorway.

My last concern would be our power on 160m. Now that all shipping and beacons have left the band I see no reason for us not to have 400 watts or transceiver power for the entire band.

I would urge everyone to be a member of the RSGB.

My First Amplifier

My first building project was a Cornishman SSB transmitter, based on the same idea as a Swan 350 transceiver using a switched VFO to get on to each amateur radio band and using a single 6146 in the PA. I finished it and then tested it while still being a Short Wave Listener. A local amateur came back to give me a report, I could not believe the box I had made was actually transmitting! - I fell over with astonishment! I put it away until I got my licence! While I was a Short Wave Listener I thought I would require a bit more than just 50 watts.

I was given a half completed G2DAF amplifier using 2 x 4-125. At the time I did not understand the great idea of rectifying some of the drive power to feed the screen to switch the valve on and off. I never did get it working successfully but I used the parts for other projects. At 17 I did not know much about valve grids and screen voltages. At my Hull Radio Club no-one used more than 50 watts CW/AM - the old timers would waffle away all day on their homemade transmitters or their 10 watt input Codar AT 5. After passing the RAE in 1970 I went straight into building a proper amplifier as my main interest was working DX on both CW and SSB.

In 1970 I made an amplifier using 4 x TT21, it gave 400 watts when driven from my newly bought National NC200 transceiver; this was replaced later with a Yaesu FT 200. I still had to build the PSU for it as funds would not stretch that far.

My next amplifier was built on the lines of one published in the ARRL handbook. My friend, John G4FCT and I built identical amplifiers using 2 x 4-400 as they were a common valve at the time and easy to get hold of. I made the box and all the internal compartments, I acquired an ex-Navy oil filled transformer and with a voltage doubler it gave 2000 volts DC. A really scary high voltage at the time! In many ways I copied the layout of the TL922. I admired the TL922 and still do, my amplifier gave me 800 watts out. I had got the bug for making amplifiers and getting them working. Who would have thought, 35 years later, I would have built over 2000 amplifiers for many applications as well as Ham Radio?

This photo shows one of my last aerials I made at Field Head before I moved house to the village of Walkington. It is a 3 el wide-spaced tri-band Quad for 15m, 17m and 20m at 70 feet at the boom.

Dyslexia and a 20 deg squint.

From my teenage years I realised that there was something strange about me compared with most other kids! I hated School though I never missed a single day. I was an only child and did terribly at school. I was not entered into any school exams and was told it would not be worth entering as it would be a waste of time. However, I would take almost anything to bits including tape recorders, drills, toasters, motors etc to see how they work and at the local Radio Club I had the nickname of Twiddly Pete.

I started school in 1955 - a place I wish to forget. Paisley Street School in Hull was a typical old pre-war school, with no gym, no lab and no playing field! I was held there until the age of 15. My headmaster Mr Kirby was pleased to see me go and his final words were, 'I would not bother to apply for a job son, you are a slow learner'. In those days dyslexia was not a condition widely recognised so I was put into the 'slow' stream, though all I wanted to do was design and make things. I wish he could have seen my Company, Linear Amp UK, 40 years on! I did have one great break at the age of 15! My father was working for ABP, Associated British Ports, as a joiner with a very good reputation. He managed to get me an interview for an electrical apprenticeship, which I got. I learnt many skills including joinery, welding, machine turning as well everything from Mains 230 volts to 11 kV and 3 phase electricity. I worked on ships and large cranes. While working on the docks I watched, with envy, ships coming in and going to all the corners of the world. I left ABP at the age of 21 and joined Court Line Shipping as an Electrical Engineer. My first ship was the Halcyon Wave a 32,000 ton tanker, which I joined in Durban South Africa as Chief Electrical Engineer. I travelled all around Africa, North America and the Far East as well as Europe. The Company collapsed in 1972 and I joined a very prestigious company called Union Castle which carried both passengers and cargo. The passengers were going to and from South Africa. The outward cargo was car parts and trucks and the return cargo was bringing citrus fruit back to the UK. A 6 week round trip from Southampton to the Cape - I was reborn and loved life. I was also allowed to operate maritime mobile so G3ZRS/MM was activated.

G3ZRS/MM

In the 1970s the government thought hobbies on board ships would be a good idea to help boredom on long journeys. I applied for a Maritime Mobile licence which was not easy to get but I did get it and used it with great joy from all over the world. I had to get the permission from my Shipping Company as well as the Radio Company. In my case Union Castle owned its own radio equipment. I also had to be checked by the Post Office and had to gain the permission of the Captain. This procedure had to be done every time I changed to another ship, which I did several times. I joined the MV Clan Robertson, a very similar ship to the Good Hope Castle which carried chilled fruit from South Africa to Europe and general cargo back to Africa. I had many contacts from my FT101E on board the Clan Robertson. It was great to talk back home from all parts of the world to my friends in the UK, mainly G3OHT, G3LZQ. A friend who was born in Lancashire and emigrated to Cape Town, Jim ZS1JJ, spoke to me almost everyday of my Maritime Mobile days. The aerial I used was a Hy Gain 12AVQ mounted on the handrail on top of the ships bridge, some 100 ft above the sea water. It was like using a beam and an amplifier.

I was 60 before I read my first book

I had never read a book but I was loaned one called THE GIFT OF DYSLEXIA by my friend Val Gardner. This was the first book I had read from cover to cover as the spacing was Dyslexia friendly. The book print is spaced for people with the problem, making it possible to read more easily. After reading it, I knew this was my big problem. I am mechanically minded with an inquisitive mind, sometimes shy and difficult to communicate with. People such as Winston Churchill, Jackie Stewart, Sir Richard Branson, Jamie Oliver, Albert Einstein, Bill Gates, Alexandra Graham Bell, Thomas Eddystone, Michael Faraday and many other famous people all have Dyslexia. I would not have thought when I left school with such a bad report that I could go on and run my own company and design amplifiers for the next 30 years with sales all over the world? This included the design of amplifiers for Universities, the MOD and many other companies doing research.

Yet another problem, a 20 degree Squint!

At the age of 56 my reading was still a problem. Whenever I tried to read a book, after the first page I would see double, the pages would be 20 deg apart. For a short time my eye muscles would correct the problem, but quickly my eye muscles would start to hurt and then I would stumble over the words. I went to several opticians without one of them finding the problem. Eventually, Dave Potter, a local optician and radio amateur friend diagnosed my problem. I was admitted to Hull Eye Hospital and a operation to correct my eye squint was performed in 2007. They cut the muscle in one side of my eye and added a piece in the other. I now have near perfect parallel sight. I still have the Dyslexia problem but at least I can now type and read better.

G3ZRS and Morse code

Morse has always been a passion and working DX very rewarding. I remember my first contact with W1BB, VK6HD and others on 160m and I operated CW Field Day for many years but needed help on receiving. My good friend G3LZQ who is a wizard on CW paired up with me as his receiving is far better than mine but my Morse sending with a paddle has always been good. I could not understand why I could read a callsign at 25 wpm but struggle if I was asked a question, or other information. The more I learn about Dyslexia explains my short term memory problem hence my problem with Morse.

My move to Walkington and new interests

When Gwen and I sold Linear Amp UK we also sold our house at the same time. We moved to the village of Walkington where Gwen had lived earlier with her parents. We have both got involved with our local dramatic group, with my dyslexia I would not attempt a part, but I do enjoy the technical side and help with the sound and effects. Gwen last Christmas was the director of the village Panto and is on course to do Peter Pan this coming Christmas.

My new QTH in Walkington showing my 2 el yagi on a 45 ft mast mounted on the back of my house. A 3el 6m yagi is also on the same boom. A trapped dipole just fits nicely in between the mast and the front and back trees. I also have a 180 ft inverted L for 160m. It comes from the mast then to the right to a nice neighbour's roof top.

G3ZRS Walkington 2014

At the age of 62 I started running every Saturday morning. What a joy! With over 450 other runners we do a 5km circuit around the fabulous East Park in Hull. I have now got my red 50th run T-shirt and working on getting my black 100th run T-shirt later this year. Parkrun is an international weekly event, a timed run with friends. I am sure there will be one near to you, do look it up www.parkrun.org.uk. I even ran in Poland twice at the parkrun in Poznan. I wish I had started much earlier in life but the chance was just not there. I am giving golf a rest at the moment but I do hope later this year when this book is finished to start again. I do find golf frustrating as a person who has to do everything well, but sometimes can't! I do play with Gwen from time to time. Last year Gwen was chosen to become Vice Captain at Driffield golf club and now this year crowned Lady Captain for the next 12 months. We both play badminton each week and I also do a long walk each week with my friend Val as well as a bit of swimming. I am doing more cycling and last year with Val completed the TPT Trans Pennine Trail, all of 215 miles. We were un-supported so we carried everything on our bikes between Southport in Lancashire and Hornsea on the Yorkshire Coast. Health is so important, playing radio is good but you must get out and do regular exercise to keep in shape and healthy.

The Passing of Vera Williams

I must finally mention my mother-in-law Vera Williams, Gwen's Mum, a wonderful person who passed away January 2014 at the age of 99. Born in 1914 and lived until 2014. Vera, for the last 20 years, had helped to look after the house and fed all our staff at lunch times, so giving Gwen and I time to concentrate on our business. All of our customers who visited Linear Amp UK will remember her as a very happy and friendly person.

Amateur Radio Products made by Heatherlite and Linear Amp UK

Heatherlite Mobile Microphone for hands-free Driving.

Mobilite   Mobile Microphone, hands free driving

First Explorer 2x 4-400   (Explorer 1000)

2m Explorer 4CX250b or 4CX350b

Later Explorer 2 x 3-500Z (Explorer 1200)

Hunter 3-500

Ranger 811H   4x 811

2m Discovery 3CX800A7

6m Discovery 3CX800A7

2m Discovery GS31b

2m Discovery GS35b

6m Discovery GS35b

70 cm Discovery GS31b

70 cm Discovery GS35b

220 MHz Discovery GS31b export to USA

Challenger 1   2x 3CX800A7

Challenger 2   3CX1500A7

Challenger 3   GS35b

Challenger 4   3CX1500A7 and GS35b

Invader   3 x 3CX800A7 for export 3 or 4 kW output PEP

SPC Tuner 1500 watts

Balanced Link couple ATU 1000 watts

Commercial amplifiers built by Linear Amp UK Ltd.

5 x 3 kW NMR medical amplifiers for Kings College London, 3 to 15 MHz

1 x 220 MHz Amplifier for Durham University

1 x 29 MHz RF Spluttering for Sheffield University

6 x 450 MHz, 1.5 KW amplifiers for the University of Montreal

18 x 5.5 kW continuous carrier amplifiers for commercial Meteor Scatter 55 to 60 MHz, these were portable units on a lorry for emergency communication use

1 x 20 kW pulse amplifier for 400 kHz earth sampling for oil and Gas, Hershel Smith laboratory, Cambridge.

The first airport colour NMR scanner, to detect objects in suit cases for Marconi research

1 x 20 kW, 400 KHz amplifier for scanning lorries at Dover looking for people and drugs.

6 x UHF link amplifiers for Satellite Communications, Pakistan

1 kW amplifier for 60 MHz, used for laser experimentation Hull

10 x 1 kW amplifiers for hospital medical use 13.56 MHz

20 x 1500watt 2m amplifiers for Dubai and the Gulf

Chapter 2

Valves

History of Valves

How a valve works

How an amplifier works

VALVES.... Who invented the valve

The history of how we started using the thermionic valves

Sir John Ambrose Fleming
(1849 – 1945)

Lee De Forest
(1873 – 1961)

Guglielmo Marconi
(1874 – 1937)

Some interesting dates in the production of RF valves, and the first Radio signals

1859 A. C. Cossor Ltd was formed making scientific glassware
1860 The first carbon light bulb was invented by Joseph Swan
1875 American G.R Carey invented the photo-electric cell.
1876 Alexander Graham Bell invents and demonstrates the Telephone
1878 Joseph Swan patents his incandescent light bulb.
1879 Thomas A Edison invents and demonstrates the vacuum carbon filament incandescent lamp.
1889 The British General Electric Company was formed
1899 The British Westinghouse Company was formed
1901 Guglielmo Marconi transmits the first signal from England to North America.
1904 Sir John Ambrose Fleming invented the two electrode radio-rectifier or thermionic valve
1906 Lee de Forest invented the Audion, the first grid triode
1907 Henry Round of the Marconi Company discovers the light-emitting diode
1914 AEG Telefunken develops valves for the Marconi Company
1915 Edison and Swan co-produce valves for Marconi Co
1920 Capt S.R Mullard forms the Mullard Radio Valve Company
1920 RCA began their first commercial electron tube manufacturing.
1921 Albert Hull invented the magnetron vacuum tube
1922 18th October the BBC Company was formed
1922 14th November the BBC went on the air from London using the call 2LO
1922 Time Signals first transmitted from Greenwich and Big Ben in London
1926 Hull and Williams co–invented the tetrode vacuum tube

The early days of radio were fraught with lawsuits as many inventors were working on similar projects at the same time. Their lives became intermingled which in turn led them to go their own ways to try to improve on what had gone before.

JOHN AMBROSE FLEMING

Fleming studied physics at University College, London and then chemistry at the Royal College of Science. He taught for a while at Cheltenham College as Science Master but all the time continuing his own scientific research. At 27 he secured a grant to again study, this time at Cambridge University gaining a First Class Honours degree in chemistry and physics, eventually being appointed the first Professor of Physics and Mathematics at the University of Nottingham. He left this post after a year to work for the Edison Electrical Light Company as an advisor for two years, then going back to University College, London as the Chair of Electrical Technology. In 1899 Fleming became Scientific Advisor to the Marconi Company where he designed the power plant at Poldhu which was instrumental in Marconi making his first transatlantic transmission. Fleming continued his research and in 1904 he invented the vacuum diode or thermionic valve, as a rectifying diode it was probably one of the most important inventions of its time. He patented this Thermionic Valve in Great Britain and also as the Fleming valve in the US, though this was later invalidated due to a technicality. The Fleming diode was used extensively for decades in radio receivers and radar and without doubt helped greatly in communications in World War II. Ambrose continued to work in many fields of electronics and radio and retired, aged 77, from University College. He was knighted in 1929 and still remained active in his support of the new age of television becoming the Television Society's first president.

GUGLIELMO MARCONI

Marconi was not himself an inventor but a very accomplished experimentalist. He became very interested in wireless telegraphy, studying at Bologna University under Augusto Righi, who had done much research into the discoveries of 'radio waves' by Hertz. In his attic, Marconi built equipment to demonstrate that messages could be sent by radio rather than by wire. Many people had tried similar work before without success. Marconi put together and improved many of the existing facts in such a way that it did work. At first he could only transmit over very short distances but by changing antennas he increased the distance. Much of his work was carried out in the UK where he received more support from the British Post Office. He established a high-power transmitting station at Poldhu, Cornwall and in 1901 claimed the first transatlantic transmission to St. John's, Newfoundland. He continued to work on transatlantic transmissions setting up stations in North America, at Glace Bay, Nova Scotia and South Wellfleet, Massachusetts. This latter station sent a message from President Roosevelt to King Edward VII in 1903, thus firmly establishing the benefits of wireless telegraphy. This station was also one of the first to receive distress signals from the RMS Titanic, making it possible to save many lives that would have otherwise been lost. Marconi continued to improve these forms of communication, though not without accusations from others for 'poaching' their ideas.

LEE DE FOREST

De Forest was born in Iowa, USA in 1873 and became interested in science at an early age. At Yale University he became very interested in radio waves and the way they could be used and when he heard of Marconi's success he built his own wireless setups. While Marconi used a coherer detector, which used metal filings, De Forest invented the responder using liquid electrolyte, which made a much more sensitive detector. With a financial backer he formed his own Telegraph Company selling his devices to the American Forces while at the same time carrying out public demonstrations all around the country. In 1903 he improved his responder but was then sued by Reginald Fessenden who had already patented a similar device; this resulted in De Forest being prevented from manufacturing his detector. At the same time he discovered that his company directors were embezzling money from him so he took his patents and left.

While all this was happening De Forest invented something which would prove to be far more important. He had been working with Fleming's diode and found that by adding a wire grid as a third element he could amplify weak signals far better. He named this device the Audion and patented it in 1906. Although he named it the Audion we know it better as the Triode and from that day to this it has been central to most of the amplifiers we know today. The Audion not only could amplify weak signals but it was also possible to modulate, or impose other signals on the radio carrier. These signals could be voice, music or any other audio, which made a massive step forward in broadcast terms. He again found backers to start a new company, the De Forest Radio Telephone Company and carried out public demonstrations of his broadcasts. He even used his system to transmit Enrico Caruso singing. His business was doing well with more contracts with the American Navy but his success was short-lived. His ability as an inventor was far better than his skill at choosing his business partners as they were all arrested for fraud. A few years after he had invented the Audion he discovered that it also had another property – as an oscillator. He applied for a patent, only to find that Edwin Armstrong had already filed a patent for a similar circuit. Initially, Armstrong won the lawsuit though after a 20 year battle the result was overturned. In many circles, even to this day, this was thought to be the wrong result and that Armstrong was the rightful inventor. None of this really mattered to De Forest because he had become very disenchanted with the whole radio business and turned his interest to putting sound on to motion pictures.

EDWIN HOWARD ARMSTRONG

Armstrong developed the regenerative circuit as a wireless receiving system and patented it in 1914. Lee De Forest claimed it was his invention and sued Armstrong. The litigation went on for years with De Forest emerging as the victor though it is widely thought that it was the wrong decision. During the years of the litigation Armstrong continued to work in the field of radio and invented FM broadcast which completely re-thought the way a signal could be modulated. At the time AM (amplitude modulation) was being used widely but Armstrong demonstrated that wide-band FM (frequency modulation) gave a clearer sound and was static-free. He battled for most of his life against authorities and bigger companies until eventually he committed suicide, depressed and penniless. After his death, his wife took up the fight and eventually the courts found in his favour over several of the patents. It was too late for Armstrong but at least it put history straight.

How does the valve work?

I think to most Amateurs the valve is similar to the average person thinking of how a television works! It works, but really do I need to know why. In fact, it is very much like an ordinary light bulb that emits light and has an electron field. You put voltage on the filament and it lights and gets hot. In a valve the cathode emits electrons and you need to control the flow of electrons between the anode and cathode, just like using a tap in the house to control the flow of water. By turning a potentiometer and adjusting the grid bias you have full control of the electron flow.

The class you want to put the valve in depends on how much bias you feed to the grid 1. The higher the negative grid bias voltage, the closer to Class C operation and zero standing current.

A controlled lower bias voltage puts it into a linear operation or AB1. For example, a valve running in class C, where the valve is biased off to zero current will give you the most efficiency - this can be used for CW or for Amplitude Modulation.

An SSB amplifier needs to run in AB1 or AB2 where the standing current is set between 10 and 15% of the full current of the valve. Full sine wave amplification is used in CW or AM. This would cause distortion if used in the SSB mode. A standing current is set just like letting your cold tap run with 10% of cold, before you apply the full pressure of hot water. A linear amplifier has to operate on the straight part of the sine curve and this then can be called a Linear Amplifier. You lose a certain amount of efficiency, but the modulation is almost pure compared with the input drive applied.

The total current flow from cathode to anode in a triode valve is determined by the electrostatic field near the cathode. The electrostatic field is a function of Ec, the grid to cathode potential, and Eb/μ, the potential due to the anode voltage electrostatic flux penetrating between the grids. The gain of the valve, μ, or amplification factor is a characteristic of the triode which is a function of the physical size and location of the grid structure.

Power Grid tubes use the free flow of electrons in a vacuum. There is an emitting surface that is the cathode, and one or more control grids to control the flow through the valve.

The anode collects the electrons. Valves are classified according to the number of grids – a triode has one grid, a tetrode has two grids and a pentode has 3 grids.

The grid is formed like a cylindrical tube and the close-wound wires let the electrons flow through yet give control of the valve current, almost like a gate in a farmer's field letting the sheep pass through.

Basic grids are made by welding tungsten or molybdenum wire to form the cylinder. To prevent primary emission as a result of deposition from the cathode the wires are coated with gold or other proprietary compounds. When used with a thoriated tungsten cathode the emissions are very low but an oxide cathode may become contaminated by barium migration.

Phillips developed the K-grid which uses molybdenum wire which undergoes several processes ending with baking at high temperature to de-gas the molybdenum and coating with zirconium carbide and a thin film of platinum. The platinum keeps primary emissions to a minimum even at high grid temperatures.

A later development of grid construction was pioneered by Eimac using pyrolytic graphite, a crystalline form of carbon. The grid is made by coating a metal former with the pyrolytic graphite to the required thickness and then laser cutting the resultant form to resemble a conventional wire grid. The advantages are that the 'wires' can be closer spaced due to the better tolerances of laser cutting and as the whole construction is in one piece there are no welds points. The required grids are made in different sizes so that they fit one over another and fixed to the valve base.

A triode only has one control grid, a tetrode has the control grid and then another screen grid. A pentode has control, screen and suppressor grids. The anode collects the electrons by being positively charged relative to the cathode. Without this potential difference no electron flow would occur. The anode consists of a metal plate which has been darkened to assist absorption and in some later valves this has been replaced by a more rugged graphite construction. A great amount of heat is generated by the anode and to maintain the efficiency and cool running of the valve, cooling must be applied. In glass valves the anode is often connected to the glass envelope. The resultant heat has to be transferred and cooled by a flow of air. The whole of the glass envelope is evacuated so no internal radiation of heat can occur. In the case of a ceramic valve the outermost part of the valve is the anode and normally has a collar of metal fins attached to it as a radiator. This kind of anode requires forced air cooling. Larger valves use distilled cooling water with a tube wrapped around the external anode.

The Valve Anode

Basically there are two forms of Anodes, internal and external types. Internal are like the 3-500Z and 572b, and external are like the ones found in ceramic valves like the one shown below.

Inside of the Ceramic GS35b Anode

The inside of a GS35 has a convex anode, it only measures 30mm across but its large external cooler helps dissipate 1500 watts. The external cooler is like a radiator and the air is forced through its external cooling vanes. This is more efficient compared with any glass valve. The same principle applies to all American and Russian ceramic triodes and tetrodes. The reliability is vastly improved due to a low power heater wattage required. The GS35b only takes 40 watts of heater power compared with 75 watts of a 3-500Z, which is only 33% of its anode capability.

The outer white part is the ceramic case which supports the anode above the grid and cathode.

Below is an Anode of a 3-500ZG which is only supported from the top of the glass envelope.

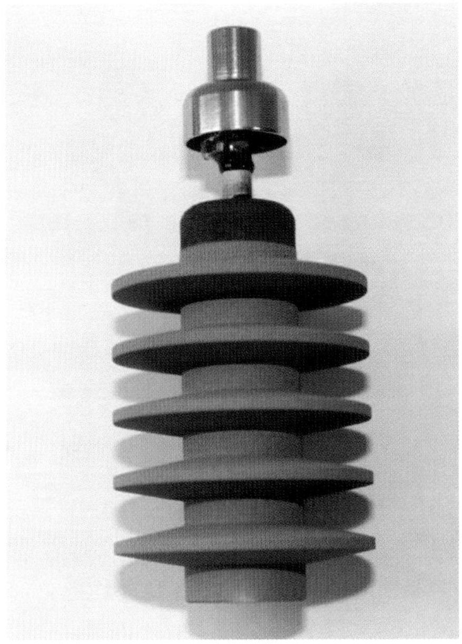

Above shows the graphite type of anode which is the ZG version.
Cooling is via convection through the vacuumed glass envelope.

Below is the anode of a 572b

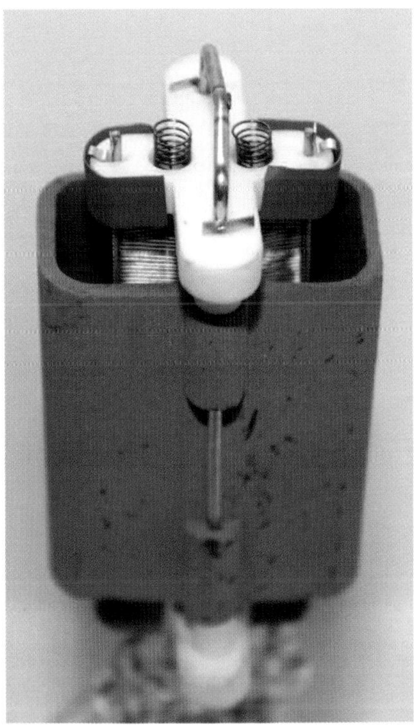

The anode of the 572b is well supported by vertical rods.

You can just see the top of the grid on the inside of the Anode tube

Direct and Indirect Heaters

Directly Heated

Advantages:

The main advantage of directly heated valves is instant warm up, just like a normal light bulb. Just hear the DX, switch on the amplifier and you have instant power.

Directly heated valves normally have a glass envelope and can be cooled easily by blowing air at the valve from a simple axial fan.

Disadvantages:

The heater power consumption of the 3-500Z is 15 amps at 5 volts, which is the equivalent to 75 watts. In an amplifier using two 3-500Z valves, which will produce 1 kW output, the power taken by the heaters is 150 watts, some 15% of the total output power.

The heater runs at white hot, compared with red on an indirectly heated valve. This is not normally a problem except that it is much more brittle and can be damaged by a sudden jolt. The only thing stopping the filament burning out is that it is inside a vacuum.

Indirectly Heated

Advantages:

For example, a 3CX800A7, which will also give 1 kW output from only one valve! The heater only takes 3 amps at 13.5 volts some 40 watts of heater power which is only 4% of the total output power. Reliability is much better, as the heater is not so fragile and under any stress, similar to a single bar electric fire.

Disadvantages:

These valves must have a warm up time of about 3 min before any drive can be applied. This can feel like an age to a Radio Amateur wanting to work an elusive DX station.
The grid must be monitored and protection fitted to prevent excessive grid current which can damage the valve grid.

Due to the construction of the valve, cooling must be achieved using forced air from a snail fan or flat pack blower.

Ceramic or Glass

In an amplifier like the Kenwood TL922 there is a pair of 3-500Z glass valves. A 3-500Z has a gain of just over 10 so you will need about 100 watt of drive for 1000 watt output. You might think that you would get twice the power from a pair of valves but the drive is divided between the two valves so the output is similar but valves run much cooler.

A typical ceramic triode like a 3CX800A7 has a gain of 40 due to its internal beam characteristic.
At 1 kilowatt output, the valve only needs 25 watts of drive, even on the 2m band! Really nice!

You might guess that the Eimac 3CX800A7 and the 3CX1500A7 are my absolute favourite valves ever made for amateur radio purposes. You would be correct! Unfortunately Eimac prices have gone through the roof, and as a one off purchase they are expensive tubes to buy.

Good news is that the Chinese are now making the 3CX1500A7 at a much more affordable price. I tried using them in the mid 1990s and the early ones were not very reliable, but for the last 15 years, the quality and reliability of the Chinese valve has been superb.

The Cathode

There are two main forms of cathode emitters.

Oxide Cathode

The filament is typically made from nickel and coated with barium and strontium oxide. A mixture of barium and strontium carbonates are suspended in a binder material and coated onto the nickel. During the vacuum processing of the valves they are baked at high temperature where the binder material is burnt off and the carbonates are changed to oxides. After this process, the cathode is ready to emit electrons. The oxide cathode is capable of high emission at a relatively low operating temperature. However, oxide cathodes deteriorate due to ion bombardment and evaporation. The recommended heater voltage must not be exceeded because higher emission will not be achieved but the life of the valve will be greatly shortened. The operating temperature is 700 – 800°C and may be directly or indirectly heated.

Thoriated Tungsten

Here the thorium is added to the tungsten in the form of thorium oxide during the making of the tungsten wire. When the valve undergoes the vacuum pumping the thorium migrates to the surface of the filament and so increases the ion emission 1000 times. The filament is also carburised which reduces the effect of ion bombardment and so increases the life of the valve. The operating temperature is 1600 - 1800°C and is directly heated. The filament can be made in three different configurations – spiral, bar and mesh.

The Valve Grid

The valve grid is the most complex part of any valve and controls its operation. In basic glass valves the grid is wound around posts surrounding the cathode, in more complex valves, the grid is a tightly wound cylinder of fine wire to let the electrons pass through it to get to the anode.

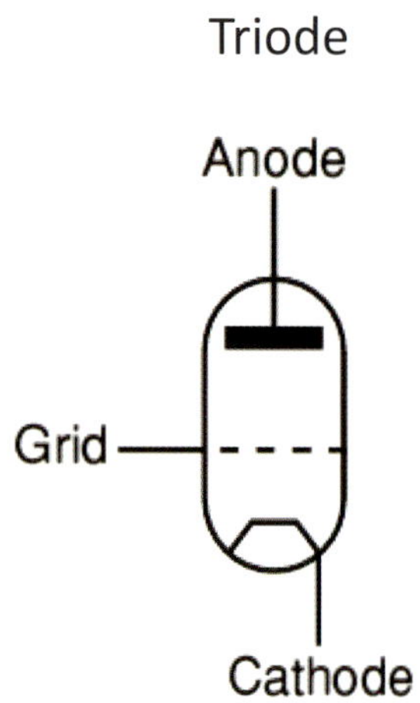

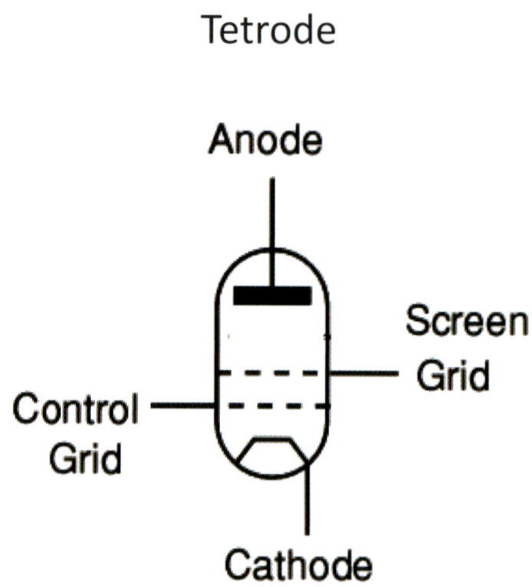

Inside of a GS35b triode showing the robust Grid construction

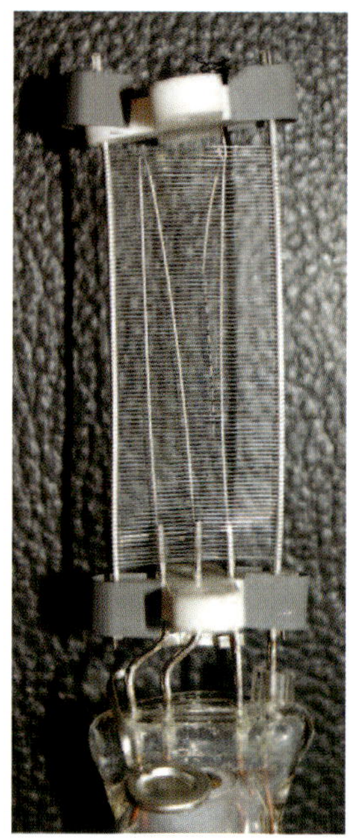

572b grid

Below shows the grid of the popular 3-500ZG

You can see the cylindrical form of the grid, also the cathode/heater in the centre.

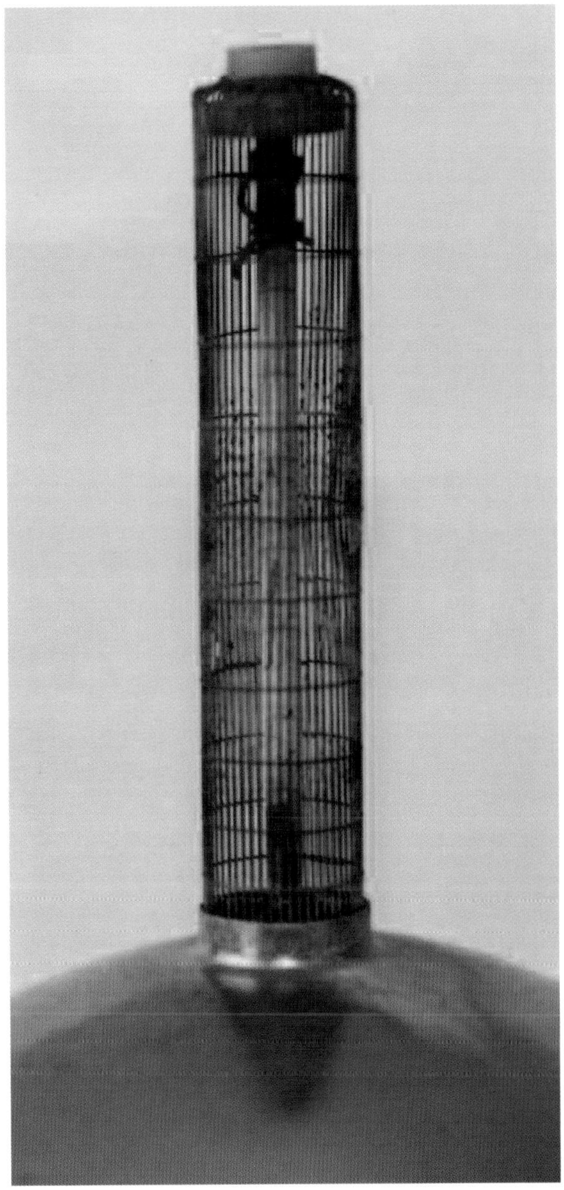

3-500Z Grid

This is a grid of an old 3- 500Z, well used for many years!

I have broken the glass envelope to remove the glass and anode to expose the vertical grid.

How HF amplifiers work and their critical parts

An amplifier increases the power of your transmitter from x 5 to up to x 100 depending on the valve used, this depends on whether the valve is a triode or a tetrode. This is called the gain and measured in decibels, and the power is measured in watts. This can be an alternating signal from Hz to GHz or audio up to the microwave frequencies.

The Fuel

The amplifier needs fuel, which must be in the form of a high voltage DC supply. The AC supply is rectified to DC. The DC supply voltage to the valve is usually in the range of 1000 volts to 3500 volts in most Amateur radio amplifiers. In commercial high power amplifiers which produce 10 kW or more this can be up to 10,000 volts or more.

The Drive

The drive from your radio transceiver is a nominal 50 ohms that has to be matched to the input of the amplifier. This can be compared with focusing a pair of binoculars to a subject that you are looking at. The input circuit of the amplifier matches your 50 ohm transceiver to the grid or cathode of the amplifier, delivering a load of 50 ohms. A cathode driven amplifier is nominally of low impedance and the matching is done with a simple Pi network input circuit and tuned for a minimum input SWR. A grid driven amplifier is higher impedance and can be done with a tuned circuit. This way gives a very high gain, typically of x 100. For example, at 2m a 4CX250B amplifier would give you 300 watts with only 3 watts of drive. The gain is determined by the load resistor that supplies the bias voltage to the grid. This can be as low as 50 ohms and you can drive your transceiver directly into the amplifier with no other matching. In this case the gain is at its lowest but there are these advantages.

1) The amplifier is at its most stable, unlikely to take off (self oscillate).

2) Most radio transceivers are 100 Watts O/P. A drive of about 50 watts is required to drive a GU74 (4CX800 tetrode) to a full O/P of around 800 watts CW, a gain of 12dB which is a low gain for a tetrode. This method can be used to drive two or more valves as the voltage across one valve is just the same as two or more valves in parallel. 2 x 4CX800 will give 1600 watts, a gain of almost 15dB, which is very respectable for an amplifier with no input matching circuit. The drive power is not added to the output, where a cathode driven amplifier is.

3) **Please note: -** Grid driven amplifiers are the easiest to overdrive. However much drive you give a passive grid amplifier the output will not increase when you have reached its current maximum as grid saturation will occur.

RF Components

RF Coupling / DC blocking capacitor

The RF coupling capacitor does a combination of coupling the RF from the valve to the Pi - network and blocks the high voltage DC getting to the other RF parts in the Pi network

The RF goes through a large coupling capacitor, usually around 1000pF – 2000pF, and this gives no resistance to RF, but stops the DC Voltage passing through to the aerial.

DC will not pass through a capacitor. DC can be stored in a capacitor if large enough, but cannot pass through it.

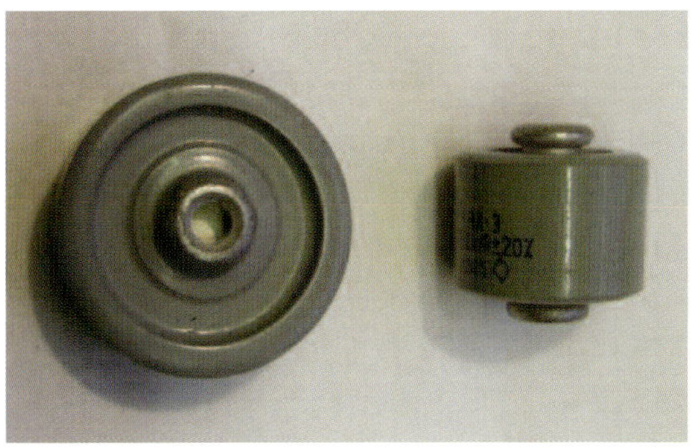

DC blocking capacitors

DC blocking capacitors are rated by voltage breakdown and also have a power rating measured in VA or wattage to tell you how many watts can be passed through it. The larger it is the more power can be passed through it. They normally use 4mm or 5mm screw unless American and that would be an American thread.

HF Anode Choke

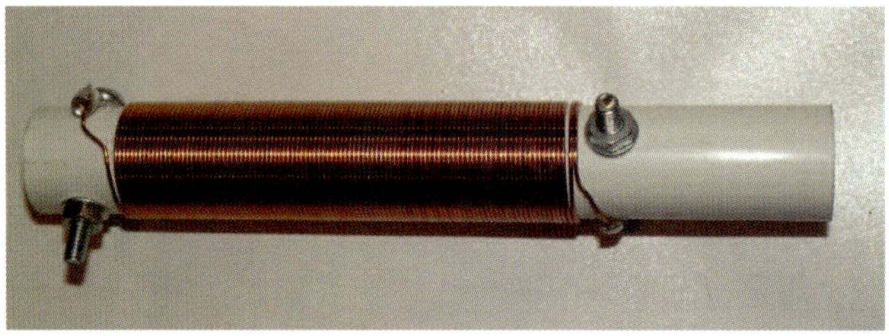

Typical Anode choke

Shown above is a typical anode RF choke on a 20mm former with 112 turns of 1mm wire.
This works well on most HF amplifiers.

The anode choke is a very important part of the amplifier, it allows the HV DC to the anode of the valve, yet stops RF getting back to the DC power supply.
Normally wound on Teflon, PTFE or a glass fibre material. The RF field is very intense within the coil so the material must be non absorbent to RF. It can always be tested by putting the former only in a microwave oven and test to see if there is any RF heating.
The coil size and diameter are critical; there must be enough inductance to stop the lowest frequency passing down to earth via the decoupling capacitor. It must not have any resonances within the amateur bands on which the amplifier is going to be used or hot spots will occur along its length. Older RF chokes often had multiple sections to avoid certain frequencies.

Description of the PA Compartment
The RF from the anode is at high (Z) impedance, this needs bringing down to a low Z. On HF there are several ways to do this, but the most common is to use a Pi network or Pi-L network, the latter giving a better harmonic rejection. The output is then tuned for maximum output using the Tune and Load controls on the front of the amplifier.

Inside a Challenger HF amplifier

Anode Decoupling Capacitor

3CX1500A7

(APC) Anode anti parasite choke

DC blocking capacitor

P A tune Capacitor

P A Coil

Anode APC (anti parasitic choke)

The anode APC choke is a small inductor fastened directly to the valve top cap. In combination with the valve internal capacitance forms the upper operating frequency. This is often called the stop frequency and is intended to stop oscillations above 30 MHz especially on the 3rd and 5th harmonics.

If the coil is squeezed together it lowers the stop frequency. If the upper limit falls in the middle of the 10m band and stops operation above 29 MHz, you can stretch the coil to increase its upper frequency limit. The APC improves stability and is usually made from 3 x 150 ohm 2 watt non inductive resistors in parallel, fitted inside the coil

The valve in this case is a 3CX1500A7

The main power house of the amplifier and the engine.
This one uses an external anode cooler and the forced air passes through the radiator to efficiently cool the anode in operation.

PA tune Capacitor and Coil

The combination of the tune capacitor, RF inductor and the load capacitor form the matching circuit. This is called the Pi Network or Pi-L circuit if a second inductor is used. It has two prime jobs to do.

The first job is to match the high impedance of the valve to the lower impedance of your aerial feeder, which is nominally 50 Ohms. Most Pi networks will cope with any impedance from 15 ohms to 150 ohms, though misleading as your SWR Bridge will show a 3:1 SWR at 150 ohms.

The second job is to stop harmonic radiation. The Pi network or Pi-L also acts as a band pass filter and you tune the circuit to the desired frequency. The circuit reduces the 3rd, 5th and 7th harmonics. Usually the 2nd and 4th harmonics are much lower and not normally a problem due to being at the wrong impedance.

Anode Decoupling Capacitors

Usually rated at around 1000pF at 3000 VDC located at the bottom of the anode RF choke and puts any stray RF that has not been stopped by the anode choke to earth, the better the anode RF choke the less RF passes to earth.

The Transceiver PTT line

QSK operation

Switching the amplifier into the transmit mode needs to be done as quickly as possible. When using Morse the first dot or dash may be omitted if the switching is too slow. Solid state amplifiers use high power PIN diodes and the more expensive valve amplifiers use a very fast vacuum relay which will switch at the same time as the Morse is being sent. When an amplifier can switch as fast as the CW is being sent the term QSK is used. This means the amplifier will allow the radio to listen between dots and dashes. There is no advantage on any voice mode. Professional maritime operators often like QSK as coast stations would work on two separate frequencies. The receive aerial would often be located at the far end of the ship to stop desensitisation. This allows the operator to hear the coast station while still sending and reply to him without any delay. If a passage was missed you could stop the sender and ask for a repeat.

This can now be done with modern transceivers and linear amplifiers using QSK.

Most modern radios now in their MENU often have a TX delay and this allows the amplifier to come into circuit before the first character is sent.

Non QSK amplifiers

If an amplifier has a slow transmit / receive relay the first character may get missed. You can send DE first before the call to make sure all the call sign is sent. On SSB a serial first digit can be clipped when using the vox circuit, just pause for a quick moment and all will come through OK. Often on newer radios a TX delay is built into the menu, this pauses for a few milliseconds before the TX is active allowing the amplifier to come in first. The PTT voltage is derived in the amplifier and sunk to earth, normally via the TX relay in the transceiver, or an open-ended collector transistor circuit. Even the newest of modern radios normally have an isolating relay to do the PTT switching. The main reason is older amplifiers have a high voltage relay coil! This is done on the Heathkit SB200 and SB220 and the Kenwood TL922 which are all over 100 volts.

How VHF/UHF amplifiers work and their Critical Parts

VHF and UHF amplifiers use a tuned anode line. This is the most common way to link the RF in the valve to the low impedance of the coaxial output. At 2m a coil can still be used for the PA tuned circuit, but a tuned line is the most efficient. This is usually a resonant tuned line and tuned to a frequency just above the band and pulled down to the frequency with a flapper capacitor. The line radiates just like an aerial, it is then coupled to the output 50 ohms either by a link-couple close to the current part of the line, just like the one below, or capacitance coupled at a high impedance point close to the valve high impedance point.

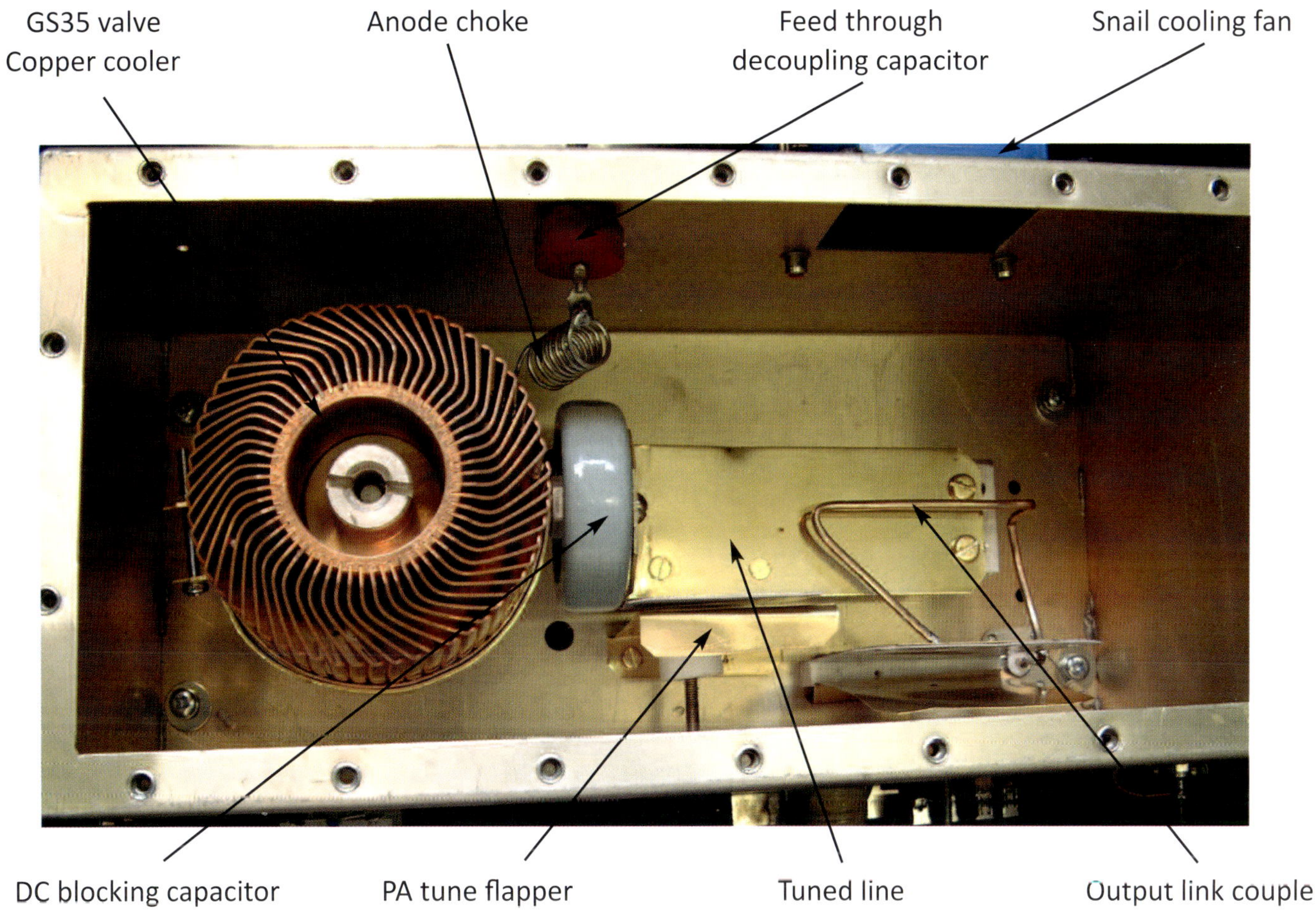

Tuned line in a Linear Amp UK 2m GS35 Discovery Amplifier

The brass tuned line is attached to the anode of the valve via a large 470 pF blocking capacitor. The end of the line is connected to the bottom of the PA box to form a ¼ wave tuned circuit.

The line resonates just above the highest frequency required. Then with a combination of the internal capacitance of the valve it will resonate just above 146 MHz. Pushing the PA tune flapper in will resonate the amplifier to the required spot in the band.

Applying Power

Switching of a valve amplifier has to be done in sequence. First the AC to the amplifier is applied. The valve heaters come on first. In most amplifiers the HT also comes on at the same time, this is not a problem as the valve will not conduct until it is warm. With indirectly heated valves like the 3CX800A7, 3CX1500A7 and the 4CX800 tetrode, you must wait 3 minutes until the heater has thoroughly warmed. If this is not done the cathode could be stripped, almost like pulling paper off a wall. This is easily done with a timer circuit and a relay in series with the transceiver PTT line.

Anode Voltage

The anode voltage is never switched on and off other than when the amplifier is switched on and off. The valve is switched on and off only by its grid or cathode.

Screen Grid 2

Tetrode amplifiers like the 4CX250, 4CX800 and 4CX1000 use the grid as the most common way of switching the valve from transmit to standby. Switching a positive regulated DC supply of 300 to 400 volts switches the valve on and off, and earthing G2 switches it off. Normally the bias voltage on the grid G1 is also increased at the same time to make sure the valve is at zero current. Grid 2 is usually earthed through a low resistance to earth on standby.

Transmit/Receive.

An amateur amplifier is no good without a bypass circuit, you cannot receive through the amplifier in circuit. This is normally done with a heavy duty double pole relay where the input and output are connected to the poles of the relay, a link is placed between the two normally made contacts, and the other two contacts go to the input drive and the output of the amplifier to the Pi net work. Alpha and other good quality amplifiers use two vacuum relays, they are quieter and give faster switching, but at more expense. A few use large PIN diodes but they can be unreliable with an accidental high SWR

Some amplifiers use a 3 pole relay, the spare pole is put in series with the cathode this switches the cathode down to earth on transmit and floats high on receive. At VHF you mostly use screened aerial switching and normally done with coaxial relays. Make sure the relays are activated before the amplifier is keyed. This stops RF on the contacts and prevents contact burning.

Chapter 3

What is RF?

WHAT IS RF?
My thoughts and comparisons

Yes, good question. What is RF? To most of us it is a method of talking to our friends on either SSB or CW, **BUT** really what is RF?

RF is energy just like AC electricity, Light or Water

AC electricity, is exactly the same as RF, it is an alternating current at a lower frequency. AC is generated as a sine wave and generated by coils passing through a magnetic field. The speed of the generator rotating sets the frequency, normally 50 or 60 cycles per second, which is then transmitted around the country at much higher voltages up to 500,000 volts. RF is generated by an oscillator/ synthesizer then amplified by a power amplifier to the power required for the radio application.

Light

Light is the nearest medium to RF, though you cannot see RF, it is at a lower frequency which is out of the range of our eyes but it acts just like a torch. With a torch you can direct a beam of light, just as you can direct the RF from a yagi or dish. This will give gain in decibels, (dB) whereas the light intensity of a torch is in lumens. You can change the frequency of RF like changing the colour lenses in a projector to get another colour or frequency.

RF radiation is like a fluorescent tube, whose light distribution is similar to a dipole - maximum in the middle and zero radiation off its ends. It is not quite the same as a dipole but for a comparison it is near enough.

Water

Water travels through a water pipe just like RF travels through coax. The amount of water that travels through the pipe is measured in gallons, the amount of RF passing through coax is measured in watts. The pressure of the water is just the same as the voltage across the end of the coax, i.e. the PD, potential difference or voltage. Water can also carry dyes or chemicals, this is like a carrier carries modulation on it.

Other applications of Radio Frequency

NMR

Nuclear Magnetic Resonance is a fantastic use of RF.
If an RF field is radiated in an enclosure like a body scanner the material within its field re-radiates at a different frequency which can be detected by a receiver. This then can be decoded to determine its property.

Ground Sampling with NMR

NMR can be transmitted by a tuned aerial wound around a tube like a torpedo, this can be lowered into the ground, sometimes up to 1 mile down. Gas and oil can be detected, in an area horizontally from the original drilling or a suitable soil structure where it may be possible to find these substances. I made the first amplifier to be used for soil sampling, one which was made in Cambridge and transmitted pulses at 400 kHz with 25kW.

Body scanners / MRI

NMR or known as MRI scanners are everyday objects in most hospitals. The RF level is controlled to a degree for just enough RF to be returned to determine tumours and bone tissue and skin tissue.

I made all the first amplifiers used at Kings College, London where the idea was originally conceived. These worked between 4 and 30 MHz with pulses of up to 3 kW

HEAT

RF is used widely in manufacturing. A parallel tuned circuit has a very large voltage across its capacitor, hence the need for a vacuum capacitor or wide spaced capacitor to cope with high voltage across it. This can be used for RF heating of wood veneers. The gap between the plates of a capacitor is called the dielectric, anything that is not an insulator will get hot if placed between the plates. The plates at a low frequency can be as big as 2 metres long and 1 metre wide so that a full sized door can be placed between. The glue can be heated to give an even heat distribution over the entire door. This is normally done at 27.12 MHz or 13.56 MHz. I have made several of these units capable of 5kW.

Microwave oven

Nearly every house has a Microwave. Of course, this is in the GHz region above what most radio amateurs use but still you can imagine the energy available. The RF passes through the item and gets absorbed as it passes through to give an even heating effect throughout.

Chapter 4

Commonly Used Valves

Glass triodes

Glass tetrodes and pentodes

High gain ceramic triodes

Russian ceramic triodes

Ceramic tetrodes

COMMONLY USED VALVES

Glass triodes

1 811 used in the Collins 30L1, Linear Amp Ranger and the Ameritron 811

2 572B used in the Yaesu FL2100, FL2100B and FL2100Z, Heathkit SB200, KW1000 and LA UK Pioneer

3 3-500Z/ZG used in the Kenwood TL922, Heathkit SB220, Linear AMP UK Explorer and Hunter 750

Glass Tetrodes and Pentodes

4 4-400 used in homemade amplifiers and marine transmitters

5 813 Pentode used in the KW500 and many home brew HF amplifiers.

High Gain Ceramic Triodes by Eimac

6 Introduction - Eimac 3CX series

7 3CX400A7/8874 used in older Alpha amplifiers, and some 2/6m amplifiers

8 3CX800A7 used in many 1500w HF amplifiers using a pair, or a single valve at VHF

9 3CX1200A7 only used in the Ameritron AL1200

10 3CX1500A7/8877 used in the Linear Amp UK Challenger, Ameritron AL 1500 and the Alpha 9500

Russian Ceramic Triodes

11 Introduction - Russian Triodes

12 GI7b/GI46 300- 400 watt valve used up to 1296 MHz in home made amplifiers

13 GS31 used in the Linear Amp UK 1 KW 2m, 4m, and 6m and 70cm models.

14 GS35b used in the Linear Amp UK HF Challenger and 2m and 6m 1500 watt amplifiers.

Ceramic Tetrodes

15 Return of the tetrode

16 4CX250B 4CX350 used in many early VHF and UHF amplifiers to give 300 – 400 watts output.

17 4CX800 or GU74 used in the Acom 1000 and 2000.

18 4CX1000A used in the Collins 30S1 and a pair in the latest Alpha 8410.

19 GS23b used in the latest OM-1500 watt tetrode amplifier

The 811, Glass Triode

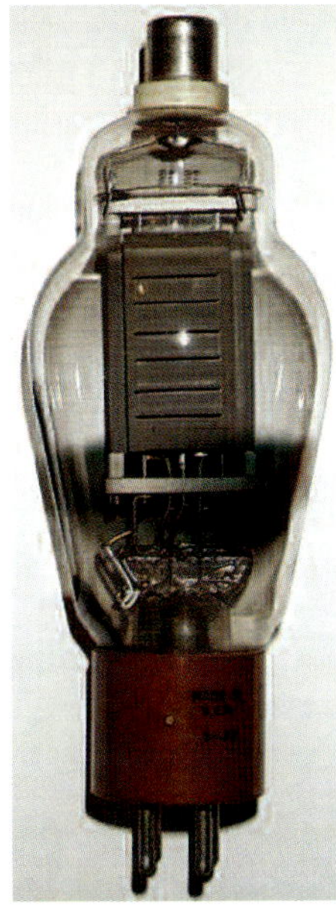

The 811 has been one of the popular valves used in Ham radio since the 1960s. The first amplifier to use the 811 was the Collins 30L1 to go along with the rest of the Collins S line. A group of 4 x 811 will give around 800 watts output. The gain is around 10 dB so you will get almost 10 x your drive power. 4 x 811 are a very stable design and easy to match the input and output. Originally made by GE and now only made in China and for a short time in Russia.

Amplifiers to use the 811
Collins 30L1
Ameritron 811
Linear Amp UK Ranger 811

Specification HT voltage range 1600 to 1800 volts

Plate Dissipation	65 watts
Heater voltage	6 .3 volts @ 4 amps volts
Gain	9 -10 dB
Base	4 pin ceramic
Grid Bias / cathode	- 3 volts
Standing current	25 mA per valve
Drive	25 watts per valve
Cooling	Low power axial fan
Max Freq	30 MHz
Typical output	200 watts
Cost	£ 25 USA$35
Pin connections	
Top cap	Anode
Pins 1 and 4	Heater
Pin 3	Grid
Pin 2	N/C

The 572b Glass Triode

The 572b Triode has been one of the most popular valves since the late 1960s. It was used in the Heathkit SB200, the Yaesu FL2100, FL2100B and FL2100Z models and also the British built KW1000. The valve can be fitted into amplifiers using the 811 i.e. the Collins 30L1 or the Linear Amp Ranger 811 as a more robust alternative having larger anode dissipation.

You will not get any more power output using the same HT but the valve will run cooler as it has a larger carbon anode. Originally made by Cetron but now made in China and badged under many different labels including Golden Dragon and Taylor. Also made in Russia and badged under the name of Svetlana.

Normally used in a pair to deliver around 700 watts with a drive of 100 watts. All designs use the valve in grounded grid driving the cathode.

Specifications

DC Plate Voltage	2400 DC
Plate Dissipation	160 watts
Bias on Grid	-2
Heater voltage	6.3 volts @ 4 amps, 25 watts
Zero signal current	45 mA
Single Plate current	250 mA
Drive power	up to 50 watts per valve
Single Tone useful O/P	300 watts
Base	4 pin Ceramic
Max freq	30 MHz
Cooling	low power axial fan
Cost	UK £45 USA $60
Top cap	Anode
Pins 1 and 4	Heater
Pin 3	Grid
Pin 2	N/C

Bottom view showing 4 x 572b or 4 x 811 bases

The 3-500 Z and 3-500ZG Glass Triode

The 3-500Z has been the powerhouse of many of the early 1kW amplifiers to give the USA legal limit of 1000 watts, now increased to 1.5kW. The classics have been the Heathkit SB220, Drake L4, L7, plus the single valve version the Drake L75, the Kenwood TL922 and the first Linear Amplifier UK, Explorer 1000. With most transceivers having a power output of 100 watts, a pair of 3-500 will comfortably deliver well over 1000W RMS output. All 3-500Z amplifiers use a grounded grid and drive the cathodes.

Articles have been written about the 3-500Z and its related amplifiers. In practice I have found the 3-500Z stable and long lasting, if not over driven. Each valve heater runs at 5 volts at 15 amps, though the cold switch-on current can be many times this, it is a good idea to fit step-start on switch on. This limits the inrush current of the valve. Cooling to the valve pins is also very important. I have known many valves desolder themselves inside their own pins due to overheat. This is due to the heater itself taking 75 watts plus the heat from the anode making the temperature very high unless adequately cooled. With an input power of around 2000 watts into the two valves, it is worth having the step-start as part of the HT circuit which helps limit the input switch-on current.

The 3-500Z was a classic Eimac tube made for many years until Eimac decided to stop production of all glass tubes. The 5 pin base has 2 pins for the heater and the other 3 pins strapped together for the grid. The tetrode equivalent, the 4-400 and 4-250 will also work in the same valve base. The heater pins and voltage are the same as the 3-500Z. Having 2 grids rather than one they both get grounded and work as a triode.

A pair of 4-400 will give around 800 watt compared with over 1kW from a pair of 3-500Z working with the same HT voltage. Of course the 4-400 is really designed to be grid driven and a very low drive of 20 watts or less would drive a pair to over 1kW output in a passive grid circuit. The 4-400 was used in many ship transmitters and commercial long distance communications.

Eimac ceased production of all glass valves, this caused quite a vacuum in the supply of the 3-500Z. To my knowledge the Russians have never made the 3-500Z. I started to use the Amperex 3-500ZG made in France. While I was producing the Explorer 1200, I was experiencing a real production problem with the Amperex valve. The reliability was poor and I got many changed FOC! I stopped buying them and this was before the Chinese ones came available. For this one reason I stopped production of my very popular Explorer 1200 amplifier.

Later the Chinese 3-500ZG came available and now I think it is the only one made in the world. There are still many thousands of 3-500Z amplifiers working every day by radio amateurs. Like other valves they come badged with different labels including Taylor and other brands, but all are made in China.

The Chinese valves when I first started using them were not that reliable. Now the quality and reliability is much improved and I would recommend them. The 500 watt dissipation is much more conservative now from a pair running 1000 watts output. The Z type would run between red and bright red (do not drive into white hot). Two 3-500ZG will run 1000 watts output with no sign of getting overheated in either ICAS or SSB operation. Cooling can be vertical through a pyrex chimney but more commonly with amateur amplifiers the whole of the glass is cooled horizontally.

A single 3-500ZG as used in the Linear Amp UK Hunter and also the Drake L75 will give 600 to 700 watts output from a single valve. A pair as used in the TL922 and the SB220 just cruises along at 1kW with no stress. This is similar to a 2000cc car doing 70 mph on the motorway compared with a 1000cc car doing the same. Some manufacturers push a single 3-500ZG to 1000 watts output, twice the anode dissipation! The tube will suffer sooner and will need replacing. After all the difference from 700 watts to 1kW is no more than 1dB which will not make the difference between working a DX station or not!

Specifications

Plate Voltage	3000 volts typical
Grid dissipation	20 watts
Plate Dissipation	500 watts
Freq Max	110 MHz
Cooling	Radiation or forced air
Heater	5.0 volts at 15 amps thoriated tungsten
Base	5 pin special SK 410
Mounting	Vertical
Gain	up to 12 dB
Typical Output	700 to 800 watts
Typical Drive	70 watts
Top cap	Anode
Pins 1 and 5	Heater
Pins 2, 3 and 4	Grid

The 4-400 Glass Tetrode

Glass tetrodes have never been used in commercial Ham Radio amplifiers to my knowledge. The 4-400 was widely used in Marine transmitters because of their high gain. The driver was either a synthesized transmitter or a crystal controlled transmitter which would only have a power output of 10 to 20 watts. Driving a tuned input circuit of a glass tetrode would give a very high gain and a good power output. You would typically get over 1000 watts from a pair of 4-400 and often known as the QY4-400 where the Q depicts the tetrode version. These valves became very popular in amateur homebrew amplifiers. There was large release of the new surplus valves coming onto the market when commercial amplifiers were being changed for the more modern ceramic tubes.

Many homemade amplifiers with the 4-400 were made, either by using grounded grid or the classic G2DAF design were you rectify some of the drive and feed it onto Grid 2. A very good design was published in the ARRL Handbook of 1982. This was the basis of my very first grounded grid HF amplifier, using no tuned input as the input is low impedance. In 1982 all radios had a valve PA so a normal Pi network would easily drive the cathodes with just a simple coupling capacitor. With 100 watts of drive you would get 800 watts of RF into your aerial - quite nice for 1982.

Specifications 2 to 50 MHz

HT Voltage	3400 DC
Input Current	270 mA
Anode Dissipation	400 watts
Screen Dissipation	35 watts
Grid Dissipation	10 watts
Drive	5 watts in Passive (Grid Driven), 60 watts cathode driven
Heater	5 volts @ 14.7 amps approx 75 watts
Bias	8.2 volts in grounded grid
Output	630 watts
Max Freq	110 MHz
Cooling	Horizontal or vertical with a SK - 406 cooling system
Type of operation	Grid driven or cathode driven grounded grid.
Base	5 Pin - Special

Top cap	Anode
Pins 1 and 5	Heater
Pins 2, 3 and 4	Grids when used in grounded grid

The 813 Glass Pentode

In the 1960s to 1980s the 813 was an iconic valve, just like the 4CX250B was for VHF. It was used widely in HF power amplifiers for amateur radio.

From memory the first transmitter in the UK to use the 813 was the famous Lab Gear LG300 which gave 100 watts AM. KW electronics made the KW 500 designed by Rowley G8KW. This was the first 400 watt PEP amplifier to use the 813 Pentode. The most common way of driving it is in grounded grid, you do not get the gain compared to running it in passive grid but the circuitry is simpler. This valve has quite a large internal capacitance which limits its upper operating frequency especially when using a pair of 813.

Specifications

HT Voltage	2500 VDC
Screen voltage	750 VDC
Grid Voltage	-95 volts
Plate Current	360 mA
Drive	60 watts in grounded grid
Power Out	650 watts
Max Freq	30 MHz
Heater/Cathode	10 volts @ 5 amps
Base	5BA
Mounting	Vertical

Eimac 3CX Series of Ceramic Triodes

Introduction

The 3CX series of valves are the best triodes ever made. Having a gain of x40, a drive of 10 watts will give an output of 400 watts and 25 watts, 1 kW. This is more what would you expect from a grid driven tetrode without all the complications of regulated screen and bias supplies. These valves have an external anode so the cooling air can be pushed directly through the metal anode. This gives a far better cooling than the old type glass triode.

In practice the 3CX series of valves are the most reliable I have known, although expensive to buy.

The design of the valve is beam focusing, beaming the electrons on to the anode like a sharp torch. These valves are extremely reliable and I used a pair of 3CX800A7 in my first Challenger I to deliver 1500 watts.

There are a few variations of the normal valves. The 3CPX version is made for pulse transmission often used in NMR (Nuclear Magnetic Resonance) MRI body scanner. I used 4 x 3CPX800A7 in an amplifier for scanning soil sampling. It delivered pulses of 20 kW at 500 kHz.

The 3CPX series have a strengthened cathode for the higher pulse operation and the base is usually a few millimetres longer. Pulses are only a few milliseconds long and each valve can operate up to 10 x its normal anode dissipation. A 3CPX800A7 will happily pulse at 8000 watt.

The group available for Ham Radio service include the 3CX400A7, 3CX800A7, 3CX1500A7. Larger ones are available for commercial service. The 3CX15000A7 will produce 29.6 kW with 8000 volts DC on the anode and a drive of 1750W.

The 3CX400A7/8874 also have a sister called the 8875, this is horizontally cooled and only requires a small amount of air in comparison. The 8875 was used in the Heathkit SB230 and Dentron MLA2500. Unfortunately these two valves are not made anymore.

The 3CX400U7, 3CX600U7, 3CX800U7, 3CX1500U7, 3CX5000U7 can been used for operation up to 1000 MHz but are more expensive.

3CX400A7/8874 Ceramic Triode 500 watts

The 3CX400A7 is the smallest of all the 3CX series valves. It has a very low profile and so an amplifier can be made a very low, small and compact for a power output of 500 watts. The Alpha 76 uses 3 x 3CX400A7 to make a very low profile neat amplifier to give 1500 watts.

The 3CX400A7 has a sister valve, the 8875, two are used in the Dentron MLA 2500 and one in the Heathkit SB230. Both valves are basically the same except for their anode coolers. The 8875 employs horizontal cooling compared with the vertical anode cooler of the 8874. Both valves must be the most difficult to purchase and replace.

I recently managed to swap an 8875 to the 8874 by altering the cooling system. The base is just the same and requires no change.

The valve works very well on 2m and 6m, but rarely used, as the 3CX800A7 is twice the power and is normally the same price or cheaper.

If you have a faulty 3CX400A7 and you cannot get hold of one, which is likely, you have a few choices. A 3CX800 will work well, but you will have to make the chimney larger and also change the heater supply to 12.8 volts. Another option is to use the Russian GI7b or GI46b. These are a similar size and have similar outputs. Both are triodes and can be used for normal grounded grid operation. The only snag, there is NO valve base, like the GS31 and GS35 they mount directly into a hole in the chassis and the grid is clamped down to earth. This is not a big job and is very well worth the modification rather than spending maybe over 1000 dollars, approximately 600 pounds/euro, on two valves. The heater on the GI7b is 12.8 volts so you would have to add a small transformer to give the 12.8 VAC supply.

A UHF version is available and useable up to 1000MHz (3CX400U7)
The base is a Johnson special called the SK 1900, 11 pin or Jedec E11-81
Recommended Chimney is the SK 696
Max freq operation 500 MHz

Specifications

Heater Voltage	6.3 volts at 3 amps
HT Anode	2200 VDC
Grid dissipation	5 watts
Cooling	Forced air
Power out	550 watts
Drive	25 watts

Valve pin connections

Top Cap	Anode
Pins 5 and 6	Heater
Pins 4, 7 and 11	Grid
Pins 1, 2,3,8,9 and 10	Cathode

3CX800A7 Ceramic Triode 1000 watts

The 3CX800A7 and the 3CX1500A7 are my all time favourite valves. Both have style and power when required, such fantastic classics of all valves. Turn the steam on and they give you fantastic power, very stable and useable to 350MHz and the 3CPX800A7 to 500MHz. The only drawback to the 3CX800A7 is its price.

When I first used the 3CX800A7 the price was only £200 each plus £25 each for the base. Unfortunately the price went up and up, and then the Chinese started to make the 3CX1500A7 for about the same price or lower!

The 3CX1500A7 is twice the power and for about same price, the base is an old type ceramic QQV06-40 valve base which can be bought for a few pounds. It made more sense to use one 3CX1500A7 instead of the 2 x 3CX800A7 that I used in the first Challengers and 2m and 6m Discoveries.

The 3CX800A7 operates at a relatively low HT voltage of 2200 volts for a valve that delivers 1000 watts. Output is very good, as low as 25 drive watts can deliver 1000 watts.

Other users of the 3CX800A7 have been Alpha, Ten-Tec Titan, Henry and Command Technology, that's just a few that I can think off!

A really great valve, but a pity about the price!

Specifications

Heater voltage	13.5 Volts at 1.5 amps, 20 watts
HT Anode	2200 VDC
Max Freq.	350MHz
Grid Dissipation	4 watts
Cooling	Forced air
Power out	1000 watts
Anode dissipation	800 watts
Drive	25 watts
Valve pin connections	
Top Cap	Anode
Pins 5 and 6	Heater
Pins 4, 7 and 11	Grid
Pins 1, 2, 3, 8, 9 and 10	Cathode

Eimac 3CX1200A7 Triode

The 3CX1200A7 is the black sheep of all the 3CX family. It is a very robust HF valve but not like the 3CX400/800 and 3CX1500A7 which will work above 2m, the 3CX1200A7 has a top frequency of only 100 MHz so it is fine up to 6m at its full rating. The 3CX1200A7 is more like a metal 3-500Z but it will deliver twice the power of a 3-500Z. The output is well over 1500 watts from one valve due to the metal anode and the efficient external cooler making it far superior to a 3-500Z. For a ceramic valve the gain is also down, where a 3CX800A7 has a gain of x 40, the 3CX1200A7 has a gain of around x 15. A 3CX800A7 which will deliver over 1000 watts only takes 3 amps or 40 watts on its heater current, but the 3CX1200 takes 21 amps at 7.5 volts some 160 watts, just about 4 times the heater current in comparison. The big advantage, like a 3-500Z, it has instant warm up but a soft start circuit is highly recommended as the inrush current could be 4 or 5 times the normal heater current at switch-on. This all depends on how much voltage sag there is from the heater transformer on switch on which is the limiting factor. I have only used the valve once to build an amplifier which ended up with my friend, G3YQA, in a Hunter amplifier which normally uses a single 3-500Z. I put a couple of more turns on the toroidal heater transformer to raise it to the 7.5 volts required. This made it into a very compact 1000 watt amplifier. It would have been a lot more if the core of the transformer had been of a larger VA rating. In many ways this is a good valve, but has never really taken off with the larger manufacturers, only Ameritron, to my knowledge, used it in their large AL 1200. As I write this I have one in for repair and the cost of a replacement 3CX1200A7 from RF parts is quoted at 1550 dollars plus import tax, 20% VAT and delivery, this is as much as the amplifier is worth. Unfortunately there is no similar valve except the 3-500Z which would really be a big downsize on a very good amplifier. The 3CX1500A7, in comparison, is a smaller valve with a much higher gain and a heater which is indirectly heated and only takes 10 amps at 5 volts some 50 watts. When a heater is directly heated it takes 3 minutes to warm up, but there is no stress on the heater. It is operated more like an electric fire compared with a light bulb as in the 3-500 or 3CX1200A7.

A good valve but do stay away due to cost and availability, unfortunately not made by the Chinese.

Eimac 3CX1200A7 Triode continued

Specifications

Plate Dissipation	1200watts	
Grid Dissipation	50 watts	
Freq for Max ratings	100MHz	
Cooling	Forced air	
Voltage	7.5 volts	
Current	21.0 amps	
Base	5 pin special	
Recommended Air system socket	SK – 410	same as a 3-500Z
Recommended Air chimney	SK – 436	
Maximum Seal & Anode Temp	250 °C	
Typical output with 80 watts of drive	1500 watts	

Valve pin connections

External Metal Cooler	Anode
Pins 2, 3 and 4	Grid pins
Pins 1 and 5	Valve heater

The 3CX1500A7/8877 Ceramic 1500 watts Triode

The 3CX1500A7 is the largest of the 3CX series used in amateur radio service. With 3500 volts on the anode it will deliver around 2500 watts with 80 watts of drive. I used the Eimac version in my Challenger 2 before the price went crazy to £700!

The 3CX800A7 and 3CX1500A7 have been my very favourite valves available for amateur radio use. Made and designed by Eimac and introduced in 1970.

The Chinese started importing their version of 3CX1500A7 around 1996. When the Chinese ones arrived for a third of the price I had to try them. In the early days the reliability was not that good, but now I think the quality is good and as reliable as the Eimac 3CX1500A7.

Besides the Challenger 2 and 4, the Ameritron AL1500, Alpha 9500 and Henry have all used the 3CX1500 with great success.

The next in size is the 3CX3000A7 and 3CX5000A7 delivering some 5 KW and 8 KW, which would be difficult to use on a normal household mains supply but I am sure someone has!

Using the 3CX1500A7 is a dream and easy to tune. It is very stable and usable up to 250 MHz. The only weakness is the grid though I have never known one get damaged with proper use and care. Considering the price of the valve, fitting a grid protection circuit is a must. The maximum grid current is 120 mA. The grid current swing can be quite large while tuning at higher powers. Delivering 1500 watts on SSB or CW the exhaust temperature is just a nice steady warm temperature. Nice for warming the shack in the long dark winter DX sessions!!

The base is the same as the old Mullard QQ- VO6-40 the number is a SK-2210

A Great Valve and nice to use!

The 3CX1500A7/8877 Ceramic 1500 watts Triode continued

Specifications

Heater voltage	5 volts @ 10 amps 50 watts
HT Voltage normal	3500 VDC
Max Freq	250MHz
Grid Dissipation	25 watts
Cooling	Forced air
Power Out	1500 to 2500 watts depending on drive and the HT supply.
Drive	65 watts

Valve pin connections

Top Cap	Anode
Pins 1 and 5 are the	Heater
Pins 2, 3, 4, 6 and 7	Cathode

Metal ring is the grid and must be earthed with fingers to earth
The large pin at 8 o clock is pin 4

Bottom view of 3CX1500A7

Introduction of Russian Triodes

My introduction to the Russian valves starts with my yearly pilgrimage to the Friedrichshafen Ham Radio show some 20 years ago. I had a booth for selling my amplifiers into Europe. At this time I was not using any Russian valves except for maybe the odd 811. After looking around the show I was astonished by how many dealers were selling these Russian valves that not many people knew about. By chance I saw a very good little book by a German Ham, Karl Weiner DJ9HO called UHF – Applikation. He had spent a lot of time making VHF/UHF amplifiers using the Russian valves. I do not think the book is sold outside of Germany but it is a host of really good information, and of course mainly in German.

I started making VHF amplifiers initially using the 4CX250B. Later I moved over to the Eimac 3CX800A7. The temptation for a good cheaper alternative was born. The Russian valves filled this gap well.

The main valves that were available were the GS35, GS31, GI7b triodes and the GU74b tetrode. The GU74b is also known as the 4CX800A for the Western market. When I first looked at the GU74b at Friedrichshafen they were only €25 each. This was before the Acom 1000/2000 and Alpha 91b and many other users latched on to them, now you could pay close to £200 for one.

Russian triodes and tetrodes are a bit of an oddity. You cannot buy them from normal trade outlets. Of course you can buy them from people selling on the Web and at some European Ham Radio Shows.

Russian Ceramic Triodes in Operation

Russian Triodes are great to use and are quite different from most Western valves. The first difference is that they do not necessarily need a valve base. The outer ring of the valve is the grid ring and hence you can fasten it directly into your chassis. Normally, most Eimac valves need an expensive de-coupling ring to ground the grid. The other two rings are the cathode / heater. It is easy to fasten a clamp round. Russian valves also have a detachable anode cooler locked on with a large locking nut. This is very useful for fitting a half-wave line on the GI7b and GS31 for 70cm.

A valve will always operate on HF frequencies. Its upper limit is always due to the internal capacitance of the valve. Most Russian triodes have a low internal capacitance and will operate happily at 70 cm and the GI7b/GI46b will also operate very well at 1296 MHz.

In practice, you will find at 70cm and 23cm you will experience thermal drift in frequency. You will need to do a small amount of re -tuning. On SSB or CW when you go to transmit you will be down by around 20%. The power will soon recover and come up to its full operating power in about 15 seconds. I do not have an answer to solve the problem other than to just accept it. I have tried extra cooling and a much thicker half-wave line to sink the heat better, but nothing makes any difference. I believe the movement is due to a stretching of the grid when hot, this changes the internal capacitance of the anode

Nearest equivalent Western valves to the Russian GS series

GI7b	400	watts	=	3CX400A7
GI46b	500	watts	=	3CX400A7
GS31b	1200	watts	=	3CX800A7

GS35b 1500 to 2500 watts depending on HT and Drive = 3CX1500A7

For a comparison the GS31b will give a similar output to the 3CX800A7, though the gain is less. For 1kW output, a GS31 will need 80 watts of drive, a gain of 12dB, whereas the 3CX800A7 will only need 25 watts drive, for a gain of 16dB, for the same output power. Grid current is also much different, a 3CX800A7 will draw a maximum of 60 mA where the GS31 is happy to take 200 - 300mA with no damage. Grid protection is highly recommended for all the ceramic triodes.

GI7b Ceramic Triode.

Base connections are the same as the GS35

This valve is the least used in Amateur Radio. If you want 300 watts at 1296 MHz the GI7b would be an ideal choice though it is not used in any commercial amplifier. They can be bought for as little as €15 each and at HF you could use as many in parallel as you would like to, so 3 would give you almost 1000 watts for the price of €45. Furthermore, the operating voltage is only around 1800VDC, they are in good supply and would make a good simple amplifier - simple to build with minimal cost for either HF, VHF or UHF operation.

Specifications at 70cm

HT Voltage	1000 to 2000 VDC
Input Current	400 mA
Drive	10 – 18 watts
Heater	12.6 volts @ 2. 5 amps
Bias required	+ 15 volts
Output	330 watts at 2kV
Max Freq	1300MHz
Type of operation	Grounded Grid, Cathode driven
Cooling	Forced air cooled with a snail fan

GI46b Ceramic triode

Base connections are the same as the GS35

The GI46b is the closest Russian valve to the Eimac 3CX400A7 which is not made any more. If you can find a 3CX400A7 they carry a very premium price. Amplifiers such as the Alpha 76A use 3 x 3CX400A7, I believe they will convert and give a similar output to the original valves. Some people think that the 4CX400 or 4CX800 are like the GI46b but they are tetrodes not triodes hence a complete change of circuitry would be required. The only main consideration would be the heater voltage and you would have to change it as the GI46b has a 12 volt heater.

The GI46b is a slightly better version of the GI7b but both are very similar in performance and output.

The GI46b would make a very good HF or VHF/UHF amplifier, 3 valves could be used to deliver 1200 watts on HF

Specifications at 70cm

HT Voltage	2000VDC
Input Current	400 mA
Drive	10 to 20 watts
Heater	12.6 volts @ 2.6 amps
Bias Required	+ 15 volts
Output	400 watts
Max Freq	1300 MHz
Type of Operation	Grounded Grid, Cathode Driven
Cooling	Forced air cooled with a snail fan.

The Russian GS31b Ceramic Triode

The GS31b is one of my favourite Russian valves. It will give 1200 watts from 160m to 2m and around 800 watts on 70cm. This is quite something for a valve that costs less than €100. It is a very robust triode and very easy to mount in a hole. For 70cm the anode cooler can be taken off and a half-wave line fastened directly between the anode and the anode cooler.

The GS31b has a gain of around 13dB, and 10dB when you get to 70cm, all depending on the amplifier build quality! I have used the GS31b in amplifiers for 2m, 6m and 70cm with great success. A single valve would also make a very easy to build 1000 watt amplifier for HF. The circuit you could use would be the same as my Challenger amplifier employing a GS35b.

Specifications at 144MHz

HT Voltage	2800 VDC
Plate Current	800 mA
Heater	12.6 volts @ 3.4 amps
Drive	100 watts
Bias required	+28 volts
Output	1200 watts
Max Freq	1000 MHz
Anode Dissipation	1000 watts
Grid Dissipation	22 watts
Type of Operation	Grounded grid, cathode driven
Cooling	Forced air cooled with a snail fan.

The Russian GS35b Ceramic Triode

The GS35b is the largest available of the Russian Triodes. Like the other valves I have just described availability is getting harder. Suppliers seem to have good stocks like QRO parts and on the Web. All mainly sent from the Ukraine and Russia, the numbers at Friedrichshafen are now lacking compared with 10 years ago.

The GS35b is quite a heavy valve at 2.8 Kg /6.2lbs. I have used it from HF up to 70cm with great success. If you want a cheap 1500 watt valve this is the one with a good specification. You should be able to pick up a GS35b for around £100/$150, compared with a 3CX1500A7/8877, a new one from RF Parts, listed at $1160 for similar specification.

You can take the main anode cooler off. The main valve body splits into two, undo the fixing nut by 2 turns, hold the large copper cooler in one hand then tap the large nut with a hammer, and it will then easily slide off. You can fasten the main body of the valve into a hole in your amplifier using 4 'L' shaped brackets. Clamp the grid ring down to the chassis of your amplifier with the 4 L brackets. The GS35b is indirectly heated so it must have approximately 3 minutes warm up time before drawing anode current. It is perfectly alright to apply the HT supply when you switch on the heaters but do not key the cathode until 3 minutes have passed.

The GS35b does not have the gain of a 3CX1500A7 but it will deliver the power when driven by a 100 watt radio. You will get 1500 watts with a 100 watt driver and HT of 3600VDC. I read, with interest, an article by a German Amateur who used a GS35b with 4000 volts on it and a drive of 200 watts to give 3000 watts on 2 metres, presumably for EME!!

The GS35b has not been made since the 1970s and all have been in cold store for many years. Buying them is strange as the Russians do not seem to have surplus sales like we do in the West. They all seem to be 'acquired' and appear at flea markets. I find them superb, especially good for 2m and 6m amplifiers. When new it is worth running the heater for an hour or two to warm the getter. This makes sure the vacuum inside the valve is 100%. When the valve is running with RF going through it the ceramic part of the valve normally glows a nice fluorescent blue and modulates as you speak on SSB and CW.

Generally the valve is good and robust, especially if you fit grid protection.

Specifications

HT Voltage	3600 VDC
Input Current	1 amp
Drive	100 watts
Anode Dissipation	1500 watts +
Grid Dissipation	26 watts
Bias required	+36 volts
Output	1500 watts
Max Freq	1000 MHz
Type of Operation	Grounded Grid, Cathode Driven
Cooling	Forced air cooled with a snail fan
Weight	2.8kg / 6.2 lbs.

The Return of the Tetrode after a gap of 40 years

If I was starting to make amplifiers now, rather than 40 years ago, I would definitely be choosing a tetrode. My life with triodes has been good, but now there is a nice choice of Russian and Chinese tetrodes. Yes, there are disadvantages, like you need a good regulated screen supply and a stable variable bias supply. It is now very easy to make the required circuits by using one of the many circuits available, or you can purchase one of Ian White, GM3SEK, Tetrode boards. It does everything you would need to run a modern Tetrode amplifier including protection, screen, bias supplies as well as a timer and all the metering and LED outputs. It comes very well documented, how to set it up and understand how it works.

My first venture into tetrodes was like many others who used the 4CX250BB to give me 300 watts on 2metres. The 4250B was actually introduced in the early 1950s, when 2m radios only gave 10 watts like the old FT221 and the Trio TS700G, so the gain of a ceramic tetrode was unbelievable. In fact, the 4CX250B could be driven to 300 watts with only 3 or 4 watts if the grid load resistor was kept to a high value. In those days the gain of the tetrode was used to its optimum.

Now the low level drive is not required any more as most VHF and HF radios all deliver 100 watts or more. A 50 ohm load resistor across the grid input makes a tetrode very stable and gives a medium gain compared with a tuned input circuit which would have been used years ago for a high gain amplifier.

With around 50 watts of drive you will have enough voltage swing across the grid to drive the amplifier to full output.

Advantages

Modern tetrode amplifiers will work on ten amateur bands. This is easily done by not needing tuned input circuits. The modern way is to drive the valve grid across a 50 ohm non inductive load resistor. This means the input of the amplifier is flat across its entire spectrum. This avoids having to make tuned input circuits for each band, which takes time winding all the individual tuned circuits.
When I first started making amplifiers we only had five HF bands where high power was allowed. We did not have high power on 160m nor did we have the WARC bands or the new 60m band. On top of all this we did not have the 6m band which is now a standard on almost all modern HF amplifiers. Having a low current heater is an advantage prolonging the heater life compared with directly heated valves.

Disadvantages

There are few disadvantages with a tetrode. You do need a more complex control board to supply the screen and bias supplies. Protection is a must as the grid parts of the valve are the most delicate. The heater is indirectly heated so you have to wait 3 minutes for the timer to come in. Having a low current heater is an advantage which helps prolong the life of the heater.

6 meters

The problem getting an HF Amplifier to work on 6m is C1, the tune capacitor, is too large! It is normally around 200 pF in size. This can be done mechanically by splitting the capacitor 80% and 20%, which gives 40pF and 160pF. The two parts are linked together for all the lower bands. Just 40pF is needed on 6m. Alternatively you could fit a 50pF fixed door knob capacitor in series with the 200pF tune capacitor. Short it out on all the lower bands. This will give a maximum capacitance of no more than 45pF.

The load capacitor C2 needs to be around 200 pF. Most HF amplifiers have a load capacitor of around 1000pF. Fortunately these are made in 3 sections and just one section can be used for 6m band to give 330pF which is ideal.

The 4CX250B Ceramic Tetrode

In 1948 Eimac introduced the 4X150A tetrode and later, in the mid 1950s, the improved 4CX250B arrived. They seem to have been around for ever for HF, VHF and UHF. It started off as a glass valve with an external anode cooler. Then it was improved into the 4CX250B which has a ceramic centre like the one above. Many companies took a licence on the Eimac 4CX250B which included Mullard, Plessey, Siemens, RCA and many others. It was such a popular valve for the high power transmitters and amplifiers.

They have been used as a single valve or up to 4 in parallel to give 1500 watts. You can easily obtain a gain of 20dB. I have seen commercial amplifiers using 10 x 4CX250B in a broadband design which gives 1000 watts with no tuning for 2 to 30Mhz.

Delivering around 300 to 400 watts with a fantastic gain, what a valve!

This is good all round valve and good for HF, VHF and UHF.

A perfect match for the Kenwood Trio TS700G and the Yaesu T221R

Most radios in the 1970s only produced about 10 watts to go to 300 – 400 watts with a single valve was fantastic.

In the 1970s a few European and Japanese companies manufactured 2m and 70cm amplifiers. These included the German Dressler D200S and the Japanese Nag using the odd 4CX350F.

I built the Heatherlite 2m Explorer using a 4CX250B to deliver 300 watts. In the 1970s and 80s 2m SSB DX was very popular before we acquired the 6m band.

I must admit I had a lot to learn about power tetrodes and amplifiers!

Many Marine Transmitters used 4 x 4CX250B to deliver 1500 watts, this has never been a popular design in Amateur Radio as a single 4CX1000B will do the same job with much less wiring.

The 4CX250B has many different versions which include the 4CX250B, 4CX250R 4CX250BM and 4CX250F and also the 4CX350B and F. Some have more rugged cathodes than others and also have different heater voltages i.e. 6V or 26V mainly for aircraft and mobile communication use.

In Europe in the late 60s the CW requirement was dropped. This created the class B G8 licence which was only VHF/UHF. Many amateurs were very happy to use 2m or 70cm with the expertise it brought. When a VHF opening came the band would be alive from top to bottom with signals from all over Europe and Scandinavia. Distances of 2000 – 3000km were easily obtainable with a good yagi and 300 watts. Of course the power helped and every radio ham in the 60s and 70s had to go to a radio rally to look for a 4CX250B and a matching base. Many articles were written about the 4CX250B and its different versions. The 4CX250B has a special base which improves stability and decouples the valve for VHF/UHF operation. I am sure many hams have bought a 4CX250B with the full intention of building an amplifier, but many were placed on a shelf like a good ornament.

The K2RIW 600 watt 70cm amplifier using 2 X 4CX250B

Finally I must mention the famous K2RIW amplifier. This amplifier was described by Ian White, GM3SEK, and John Norman, GW4FRX, in the 1995 RSGB VHF/UHF hand book.

It uses two 4CX250B to deliver around 500 to 600 watts, a great power for those days for the 70cm band. This would be enough power for serious DXing, also Earth Moon Earth communication, when feeding a good aerial array.

Employing a half-wave tuned line for the output circuit which was capacitance coupled to the output. The input is a tuned brass line which is also directly coupled by a coupling capacitor.

In those days there was not a lot of choice of valves to give high power on 2m and 70cm.

Now the availability of the Russian GS series of valves gives a good choice of valve for VHF and UHF frequencies.

Specifications HF to 500 MHz

Anode Dissipation	250 watts
Typical Output	300 to 400 watts
Drive	< 10 watts
Heater	6 volts @ 2.4amps, 15 watts
Cathode	Oxide- coated Uni potential
Operating position	any
Plate Voltage	2000 VDC
Screen Voltage	300VDC
Base	9 pin Special SK-600
Chimney	Special SK 606
Cooling	Forced air cooled e.g. snail fan.
Top Anode Cooler	Anode
Valve base centre pin	Grid 1
Pin 1	Screen/Grid 2
Pins 2, 4, 6 and 8	Cathode

The Russian GU74b/4CX800 Ceramic Tetrode

The most popular tetrode is the GU74b made around 1980 later called the 4CX800A and must be one of the most used tetrodes in the Amateur radio field. I do not know how many were made but it seems like tens of thousands, the supply never seems to dry up. They have not been made for almost 30 years. Supplies of the GU74 are now getting low and the price is going up from the 25 euro I paid at Friedrichshaven. The GU74b has been rebadged for the west as the 4CX800 to come in line with Eimac numbering similar to the Eimac 4CX1000.

Original Russian GU74b have a dull finish, almost like lead colour, but many went for re-silvering and came back looking like a new 4CX1000. This was all done by George Badger W6TC, who was ex-Eimac.

The Russian valves were made 30 years ago with the name of OTK on them, in their factory in St Petersburg, Russia. Described as a high μ tetrode, it is ideal to be grid driven using a 50 ohm input load resistor eliminating any other input circuitry like grounded grid amplifiers need. George Badger W6TC realised its potential and saw the opportunity of using the Russian valves. He created the name Svetlana to market the newly found tetrodes into the west. All the GU74b valves we use now are NOS, new old stock.

Some Amplifiers that use the GU74:-

Alpha 91β, Acom 1000 and 2000 Auto, Emtron DX1 and DX2, Alpin amplifiers

The Russian GU74b/4CX800 Ceramic Tetrode continued

Specifications HF and VHF amplifiers

Anode dissipation	800 watts
Typical output	800 to 1000 watts
Drive	60 watts into a 50 ohm load resistor
HT	2200 VDC
Screen	300 to 350 volts dissipation @ 15 watts
Max Freq	250 MHz
Warm up time	3 min
Cooling	Forced air e.g. snail fan
Heater	12.6 volts @ 3.6 amps, 45 watts

Pins 3 and 7	Heater
Pins 2, 4 and 6	Cathode
Pin 1	Grid 1
Pin 5	Screen grid 2

The Eimac 4CX1000A   Ceramic Tetrode

The Eimac 4CX1000A is a fantastic 1500 watt tetrode. A valve that was probably the most used valve in commercial high power amplifiers in the 1960 to the 80s. It was brought out by Eimac in 1957 and used in the classic Collins 30S1. I have also seen them used in VHF Police Transmitters using several frequencies from one valve to give several 200 watt channels multiplexed from one aerial.

In the UK most Hams did not consider it, with a legal limit of only 400 watts using a 4CX1000A in the 70s and 80s would have been frowned upon. We still had inspections by the 'Man from the Post Office', later the RA and now OfCom. That is a thing of the past now and everyone is left to police themselves. Many manufacturers started using Russian tetrodes like the GU74/4CX800 but now with them getting rarer and more expensive the 4CX1000A are becoming popular again. The new Alpha 8410 uses two instead of a pair of 4CX800. The Chinese make them and appear under the name Taylor (RF Parts) and other badged labels, but all are made in China, unless specified Eimac, which are still made in the USA.

I think if I was still making amplifiers, this would be my first choice. The modern idea of using a 50 ohm load resistor to feed your transmitters into is so simple and requires much less wiring. You do not have any tuned input circuit but a Pi-L network is necessary in order to keep the harmonic content to an acceptable level.

The Eimac 4CX1000A Ceramic Tetrode continued

Specifications up to 110 MHz

HT Supply	3000VDC
Screen Voltage	325VDC
Anode Dissipation	1000 watts
Screen Dissipation	12 watts
Heater	6 volts @ 9 amps, 54 watts
Max Frequency	110 MHz at full CW rating.
Mounting	Horizontal or vertical
Base	Special SK-800B
Cooling	SK-806 air chimney, with snail type fan.
Weight	0.77 kg
Drive	50 watts with 50 ohm Grid load resistor.
Typical output	1630 watts

Top Cap	Anode
Top ring	Screen
Pin 3	Grid
Pin2	Heater /Cathode
Pin 1 Bottom Ring	Heater

GS23b Ceramic Tetrode

The GS23b is the latest tetrode to be used by Western manufacturers. You would think by its number it would be a triode but it is actually a tetrode.

I first ran across the GS23b in the OM 1500. The valve also has a western number of the 4CX1600U. It has a very high specification as it can give 1000 watts on 70cm. It is also is useful at 1296 MHz with reduced ratings. It is built as a coaxial valve to keep everything symmetrical for its UHF properties.

Specifications HF - UHF

HT Voltage	3000 VDC
Screen	500 VDC
Grid	- 150VDC
Output	1500 watts
Heater	6.3V @ 6.25 amps
Anode Dissipation	1500 watts
Max Frequency	1000 MHz full rating, 1300MHz at a reduced rating
Cooling	Forced air cooling e.g. snail type
Typical gain	15 dB using a 50 ohm input load.
Base	UHF special

Chapter 5

Some Classic Valve Amplifiers

Acom

Alpha

Ameritron

Collins

Drake

Emtron

Heathkit

Linear Amp UK

OM Power

Ten-Tec Titan 425

Yaesu

Classic Valve Amplifiers

Acom Amplifiers

Nikola Mushanov 151
1330 Sofia, Bulgaria — web www.acom-bg.bom
Tel 359 2 920 96 55 — email acom@mail.orbitel.bg

Acom started out as a joint venture with Alpha, they developed the Alpha 91β. The model was built in the States first as the Alpha 8100 then they moved production to Bulgaria to make use of the cheaper Russian valves and components and re-named it the Alpha 91β

Later Acom brought out the Acom 1000 to be more in line with European power levels using a single GU74 in a normal grid driven design, a good amplifier to give 1000 watts. The price is reasonable and really good value for money, the build quality is good using mainly Russian components.

The Acom 2000 uses 2 x GU74 which will deliver 1500 watts. This is a cheaper version of the famous Alpha 87 which used all American parts compared with the Russian parts used in the Acom 2000.

After a dispute between Alpha and Acom, the partnership finished, as did the production of the Alpha 91β.

Models now include:-
Acom 1000, Acom1010, Acom 2000 auto tune.

Acom 1000

The Acom 1000 the most popular HF tetrode amplifier ever made.

With kind permission from Val LZ1VB of Acom, I show their very popular tetrode amplifier.

It has very clean design inside and out. I think this is the most popular tetrode amplifier ever made at a very sensible price.

Front Panel layout of the Acom 1000

The Acom 1000 covers 6m to 160m in seven switched bands. Like many amplifiers 10m/12m and 15m/17m use common switch positions. The LCD display toggles through the operating parameters. It will delivers up to 1000 watts output with a drive of just 50 to 60 watts. It has good protection and monitoring. The amplifier is fitted with fast vacuum relays allowing good QSK operation. The Acom 1000 uses a GU74 to deliver up to 1000 watts

These two pictures show the top of the modern Acom 1000

There is nice separation of the RF deck from the rest of the amplifier.

You will see how the Tune capacitor is split 30% and 70% to allow 6m Operation

RF Deck of the Acom 1000

Side view of the RF part of the Acom 1000

Outline circuit of the Acom 1000

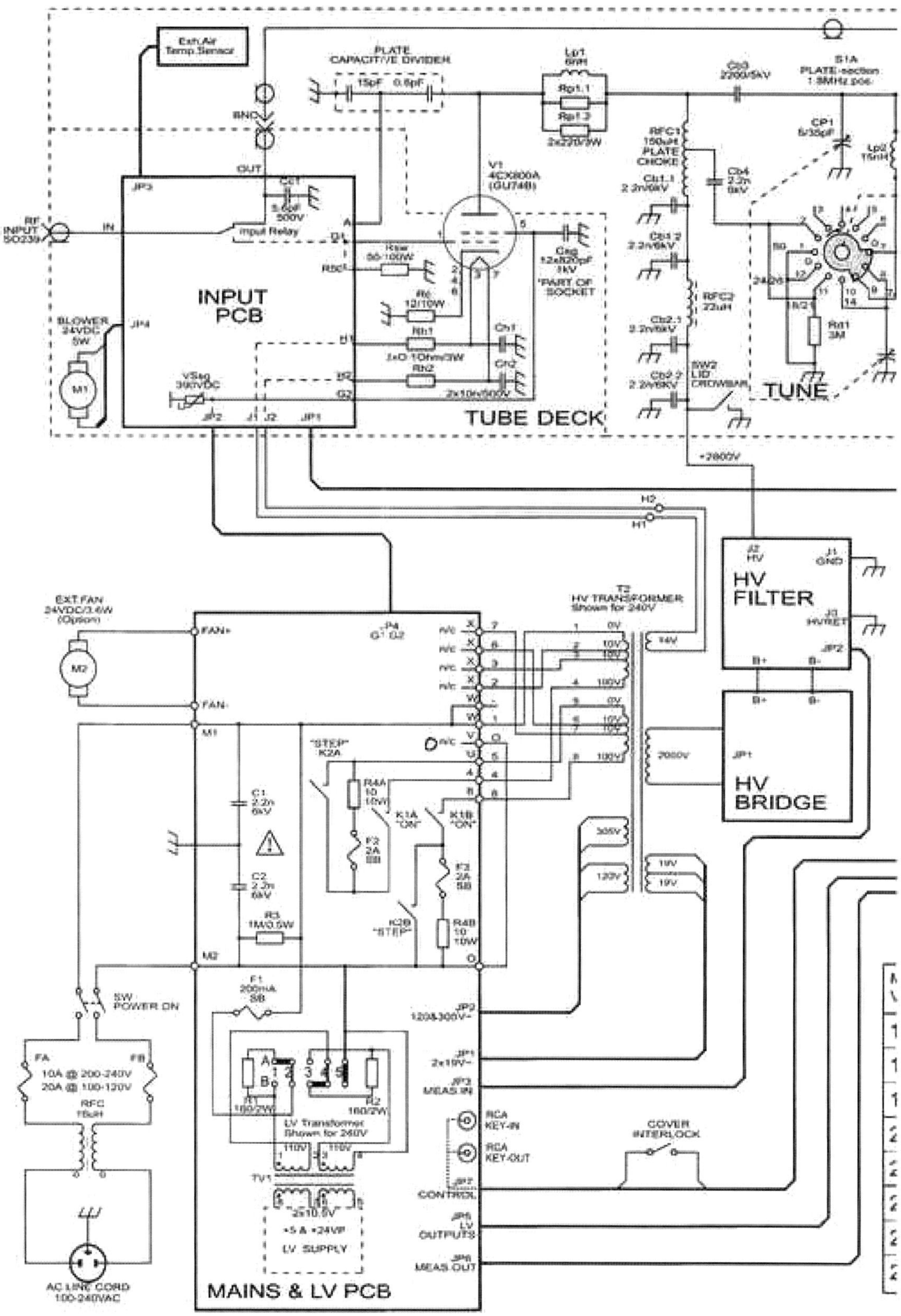

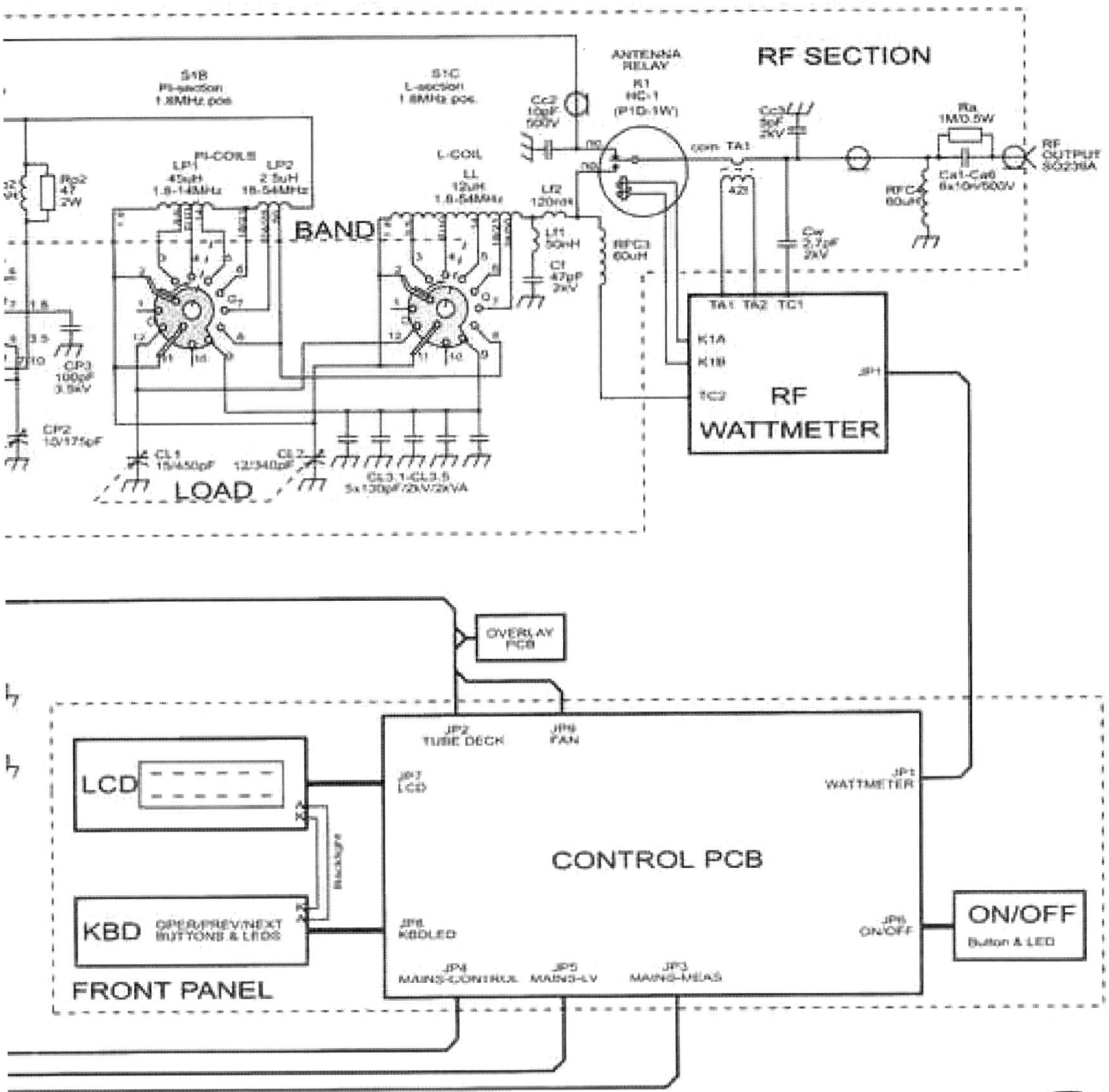

Mains Voltage Selection

Mains Voltage	LV transf.	HV transformer 8	4	U	V	W	W	X	X	X	X	Rear Panel Fuses
100V	A B 1 2 3 4 5	8	4	0	–	3	7	1	2	5	6	20A
110V	A B 1 2 3 4 5	8	4	0	–	2	6	1	3	5	7	20A
120V	A B 1 2 3 4 5	8	4	0	–	1	5	2	3	6	7	20A
200V	A B 1 2 3 4 5	8	4	7	0	3	–	1	2	5	6	10A
210V	A B 1 2 3 4 5	8	4	7	0	2	–	1	3	5	6	10A
220V	A B 1 2 3 4 5	8	4	6	0	2	–	1	3	5	7	10A
230V	A B 1 2 3 4 5	8	4	6	0	1	–	2	3	5	7	10A
240V	A B 1 2 3 4 5	8	4	5	0	1	–	2	3	6	7	10A

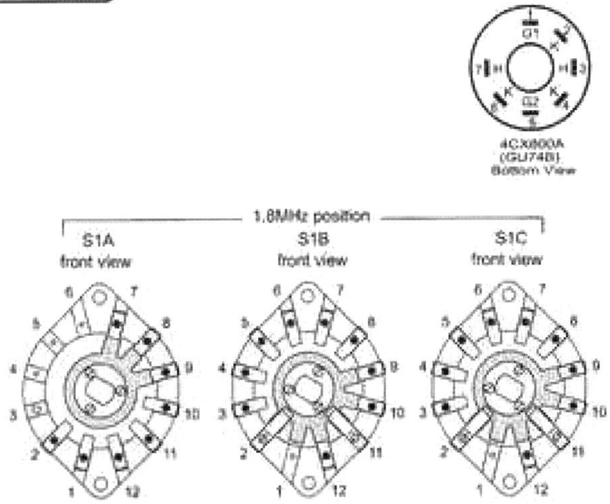

Fig. 6.1 ACOM1000 Simplified Schematic Diagram

RF Concepts and Alpha Power Amplifiers

Owned by Michael Seedman AA6DY
Vice President Gordon Harman
Based at RF Concepts 634 South Street Longmont Colorado 80501 USA
Tel 303-473-9600 web www.rfconcepts.com

Alpha was founded by Dick Erhorn W4ETO and has always made high quality amplifiers. Now known as RF Concepts, Alpha was the first company to have a fully automatic band hopping amplifier, the Alpha 87A. The 87A has been a contesters dream being able to band hop and with manual retuning. Gwen and I met Dick and his sales manager John W0UN at Friedrichshaven in 1996. John W0UN also retired and moved to Texas. Dick sold the business to Michael Seedman AA6DY.

The Alpha 87 automatic amplifier employs a frequency counter to detect the part of the band you are on. The amplifier then goes to the pre-tuned positions using stepper motors. Alpha has always used all my favourite ceramic triodes like the 3CX400A7, 3CX800A7 as well as the 3CX1500A7.

The vast number of HF amplifiers made by Alpha using ceramic valves include:-

Alpha Models

86, 89, 78, 87	using	2 x 3CX800A7
76A, 78, 374A	using	2 x 3CX400A7 or 8874
8100, 91ß,	using	4CX800, GU74
77DX,	using	2 x 3CX1500A7

RF Concepts

9500 auto using 1 x 3CX1500A7 the latest model which follows on from the 87A
8410 manual tune using 2 x 4CX1000 tetrodes the latest manual tune model

The Alpha 87A

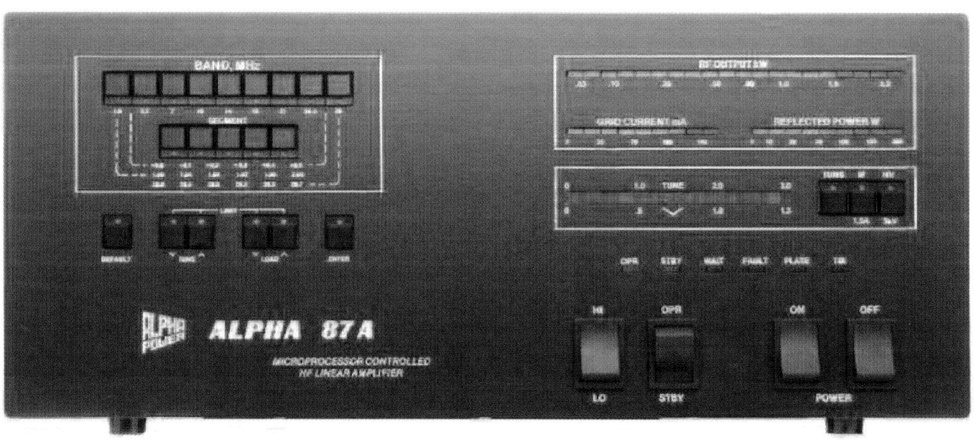

The above Alpha 87 has been the Dxers dream amplifier for many years. From its pair of 3CX800A7 it delivers 1500 watts with ease. It is auto band change and remembers the tune and load setting for each segment of each band. It has now been replaced by the RF Concepts 9500 which uses a 3CX1500A7.

The Alpha 9500 and 8410

The 9500 Auto and the 8410 manual are the latest current models from Alpha now RF Concepts. I was sent the first of each to come into Europe. The two new amplifiers are both very impressive and keep up the high standard of Alpha and RF Concepts. I arranged for the CE testing to be done with a CE test house in Essex. RN Electronics made the appropriate modifications to meet European CE standards. The new owners contacted me when they decided to introduce both models into Europe. I could have become the main distributor in the UK and Europe but I decided not to as I had just sold my premises and business and having a break from amplifiers! At the time I was thinking more of retirement. Later RF Concepts chose Martin Lynch and Son as their UK outlet and after sales service centre.

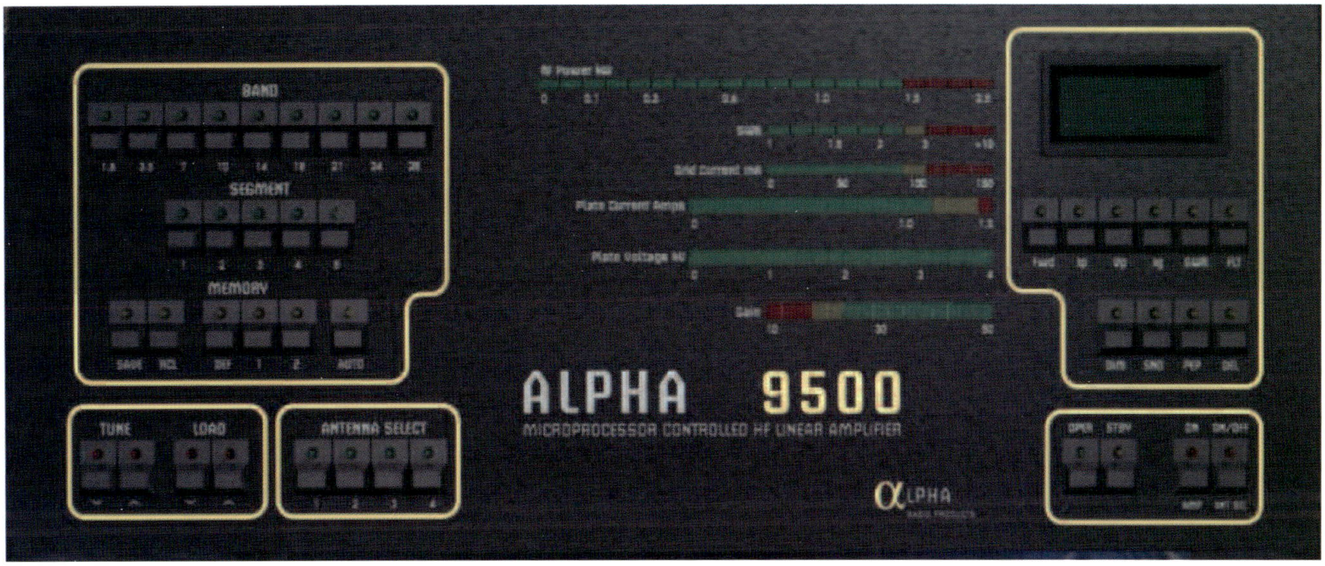

The New RF Concepts 9500 Automatic Linear Amplifier

The RF part of the 9500 Auto

The latest RF Concepts 8410 Manual tune amplifier

The impressive power house of the 8410 with its 2 x 4cx1000A

The 4CX1000 is a good choice and is still made and readily available from Eimac or Chinese outlets.

Ameritron MFJ Enterprises INC

PO Box 494 Mississippi State, MS 39762
Tel 1-800-647- 1800   web www.ameritron.com

Ameritron has the largest range of HF amplifiers and other units including ATUs, analysers, aerial switches etc in the world. MFJ has bought out a lot of the big name American Companies. MFJ has had a vision of making a vast range of Amateur Products. MFJ have not made high power VHF or UHF amplifiers. The products are good value and give good value for money. They have also just launched their latest 1200 watt solid state amplifier, similar to the Yaesu Quadra and the Icom PW1 and the Expert 1K using MRF 150.

Ameritron Amplifier range includes

AL 811	using 3 or 4 811 to give 600 or 800 watts
AL80b	using a single 3-500ZG
AL82	using 2x 3-500ZG
AL 1200	using the 3CX1200
AL 1500	using the 3CX1500A7
AL 800 H and HX	using 1 or 2 x 3CX800A7

MFJ have other companies which are Hy Gain, Mirage, Vectronics and Cushcraft. Their corporation product range is large and they probably have the largest range of products of any manufacturer.

One of many Ameritron amplifiers, this is the AL-80

Using a single 3-500Z giving up to 1000 watts 10 to 160m

The above AL80 would be my choice of medium power amplifier from Ameritron. The 3-500Z is more robust than the 811s as used in the AL811. It has nice clean lines and all components are quite easy to get to. The amplifier is quoted at giving 1000 watts but I would happier running a 3-500Z up to 750 watts, 1.5 x the anode dissipation. The above American version does not cover 10m but all other export models do, though it is quite easy to fasten on the wiring for the input tuned circuit and the PA 10m coil connection. Both the tune and load capacitors have slow motion drives making the smooth and easy.

Ameritron AL 811

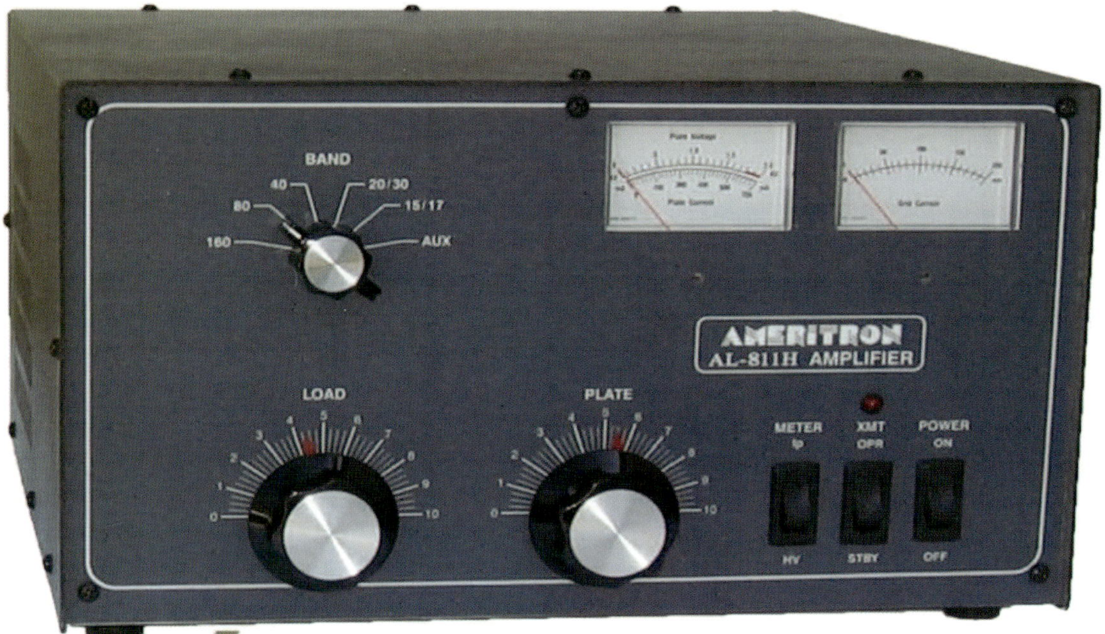

The Ameritron 811 is the most popular 811 amplifier ever made. It is available at a very reasonable price. It uses 3 or 4 x 811 in grounded grid to produce 600 or 800 watts output. The construction is very open and most components are easy to get to. The amplifier is built on a single base plate. The back panel has all the control and input circuits. The amplifier can work with 811s or 572b valves and runs at 1800 volts DC. The power supply and Pi network are suspended in front of the 4 valves on long pillars.

The Ameritron AL 1500

The Ameritron AL 1500 is the largest of the Ameritron range. It is a very substantial amplifier in every way. It uses my favourite tube, the 3CX1500A7 ceramic triode, to deliver well over 1500 watts. With this amount of anode dissipation it would run any mode at 1500 watts with ease. It is shipped with the transformer removed, installation is easy with plug connectors on the mains transformer. The 3CX1500A7 is a very reliable valve and if run properly and kept within its normal parameters will hardly ever go faulty. Original valves used are Eimac but now there is a choice of Eimac at cost of around $1500 and Chinese equivalents for around $665. I have used both and find the latest 3CX1500A7 coming from China good and reliable. With ceramic tubes it is always worth running the heater for an hour or so before you connect the HT, especially if the valve has been stored for a while.

Ameritron also make the AL1200 which uses the 3CX1200 and looks similar to the one above. A very robust valve and in many ways is like a ceramic 3-500Z. The base is the same as a 3-500Z with a heavy heater current.

The 3CX1200 is different in the fact that it is instant warm up. It has a heater that draws 21 amps at 7.5 volts. A replacement valve is not available from the Chinese as they do not make it, to my knowledge. A new Eimac one costs around $2000!! I would look into fitting a Russian GS31b or GS35b with a few metalwork changes. The heater supply would need changing to 13 volts. Normally an AL 1200 would not be worth this much on the second hand market. So it makes you think would it be worth cutting your losses and think about a new amplifier.

Collins Radio Company

Formed in 1933 by Arthur Collins in Cedar Rapids, USA, their range of equipment has been vast. He started with audio equipment and moved on to AM broadcast equipment and military communications. The main business has been mainly in Avionics, and as well as supplying the VHF and HF radio equipment on aeroplanes, they also do the in-flight entertainment found on most aeroplanes now. Collins, after financial problems in 1973, was bought out by the Rockwell International Corp and is now known as Rockwell Collins. The company has now bought out many other companies to do with the Avionics business

Rockwell Collins now employ around 20,000 (2010) people and a turnover of 4.5 billion dollars. The main company HQ is in Cedar Rapids Iowa. To Radio amateurs, owning a piece of Collins equipment was like owning a new Rolls Royce motor car. Owning any part of the famous S line equipment now is just like owning an old classic motor car! I remember tuning around the radio in the 1960s and hearing G6OF who had just bought a new complete Collins S line including the 30L1, it was outstanding speech quality, and their famous radios including the S -line and KWM2 were expensive but the best of all available at the time. I remember around 1970 you could buy a good receiver for around £ 100, a Drake R4C would be around £250 and a Collins 75S3 receiver was around £500, a complete Collins S-line with 30L1 would set you back about £2000 and when you consider you could buy 3 bedroom semi house for about £ 4000, this was a lot of money by anyone's standard.

For Amateur Radio, Collins made the 30L1 to go with the portable KWM2 transceiver, which was often sold with a carrying case that could be taken with you on a flight. The combination was often taken to remote countries as the first mode of communications.

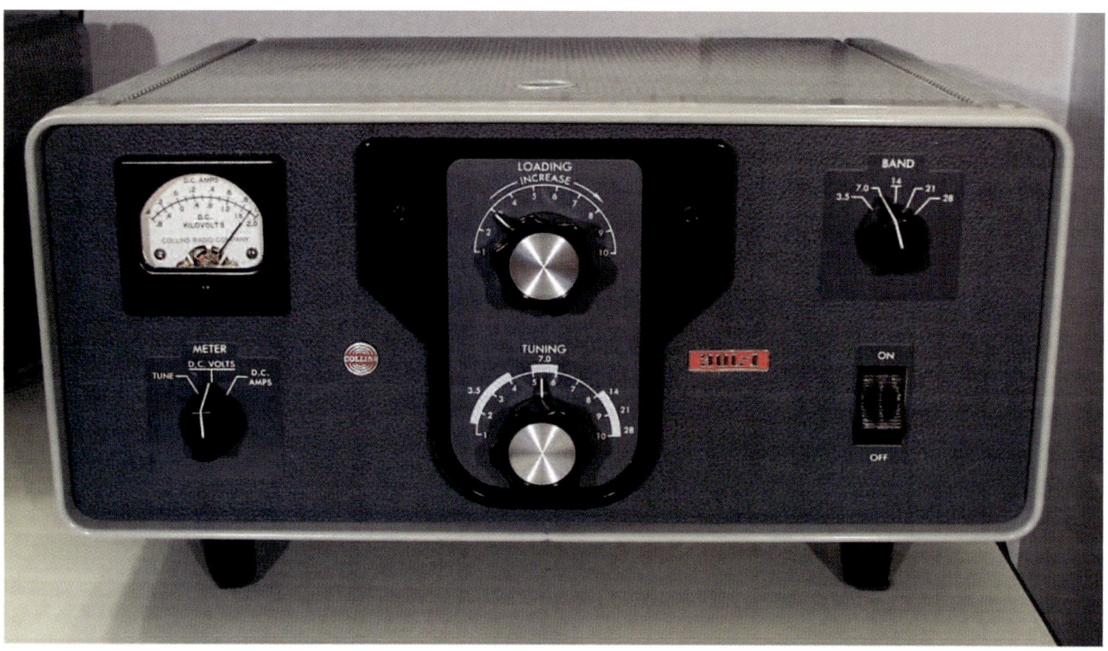

Collins made 2 classic Amateur Radio amplifiers, the 30L1 pictured above using 4x 811 giving around 800 watts, and the Base station 30S1 using the 4CX1000 to give 1000 watts

I have at some point in time repaired just about every model of amplifier ever made but the only one I have not, is the Collins 30S1 which is a large unit and hardly ever seen outside of the USA. It stands on the floor on castors so it can be moved around. It was made to cover 3 to 30 MHz so it could be used for commercial traffic as well as amateur radio. The 30S1 gives 1000 watts from a single 4CX1000. The 4CX1000 was designed by Eimac in 1957 as a very modern high power tetrode. The whole unit looks so professional and you would stand in awe looking at it. Delivering 1 kW with ease, it would look the perfect part when fitted in an airport control centre, to communicate with in and outbound flights. Collins last product made for Amateur Radio was the KWM380 but it was never a great success as the Japanese had started to make good quality radios at affordable prices with many extras fitted, which you expect as standard these days. The KWM 380 did not even have an RIT control and cost a lot of money. After this no other products were made for Amateur Radio and the company concentrated totally on commercial products, mainly in Avionics.

Collins 30 L1 is a classic grounded grid amplifier using four horizontally mounted 811. The amplifier would give around 800 watts output with 100 watts of drive. Construction was excellent as with all Collins products. The 30L1 covered 10 to 80m and came complete with an internal power supply.

Along with the matching KWM2 transceiver these were made to fit in travel suitcases and were often the first mode of communications from rare and difficult locations around the world in the 1950 and 1960s.

The 30L1 inspired me to make the Linear Amp UK Ranger 811. I added soft start and 160m along with the WARC bands. A 30-L1 was in service for many years at GB2SM the science museum in London until they replaced it with a Linear Amp UK Explorer1200.

For fixed station operation, the Collins S line comprises the Collins 32S1 Receiver, 75S3 Transmitter, speaker control unit and the 30L1 Linear Amplifier, which would have been the most expensive station in the 1960s and 1970's.

The Collins 30S1 10 to 80m 1 KW Linear Amplifier

The Collins 30S-1

For a long time this has been probably one of the most sought after amplifiers in the world and would still fetch a premium price like a rare painting. The 4CX1000 is not pushed and will run a very comfortable 1000 watts. In many ways the 30S1 is old fashioned but this adds to it beauty like a classic car or well restored steam engine.

I think if you had to choose the most classic amplifier ever made this would be the one. It just sits there like a statue. It is well laid out and very conservatively rated at 1000 watts. It uses the Eimac 4CX1000 tetrode which Eimac introduced in 1957.

Not like the 30L1, it is not intended as a portable amplifier but to be placed in a radio station either ham or commercial use and just sit there. It could be tuned to almost any frequency between 3.4 and 30 MHz.

R.L. Drake Company

Founded in 1943 by Robert L Drake, the company was based at Franklin, Ohio. Drake, just like Collins had an extremely good name for quality products, initially making low and high pass filters. After the Second World War, production started in the Amateur radio range. Between 1950- 1970 all the amateur radio products were made and are still very desirable units some 40 to 60 years later.

Amateur radio products including famous receivers like the Drake 1A, 2B and latterly the well known Drake 4 line, which was the T-4X transmitter and the R-4C receiver often called the Drake twins. The equipment was built unusually on a copper chassis. Drake made latterly, the R8 receiver and the TR7 solid state transceiver.

Drake made three amplifiers while in amateur radio production, the L4, later the L7 and a smaller model called the Drake L75. The L4 and L7 both were made to give 1000 watts from 2 x Eimac 3-500Z. The HT was quite low at 2400 volts compared with other amplifiers as all the Drake transmitters, including the TR7, would give 150 watts output.

The Drake L75 is a self contained unit with a single 3-500Z and an internal power supply. It covered 10 to 160m and delivered around 700 watts.

The L4 and L7 both had a power supply to sit on the floor with a connecting cable carrying the 2400 volts. Better keep the pets away!

Robert L Drake died in 1975 and the company was then run by his second son Peter W Drake. Drake does not produce Amateur radio equipment anymore but are involved with the manufacture of equipment for the Digital Cable TV distribution industry.

In 2012 Drake was sold out to Blonder Tongue Lab. Inc. for a price of $6.5 million, so the name and company still thrive, though not in their well known Amateur Radio market.

RL Drake and Co L-4B

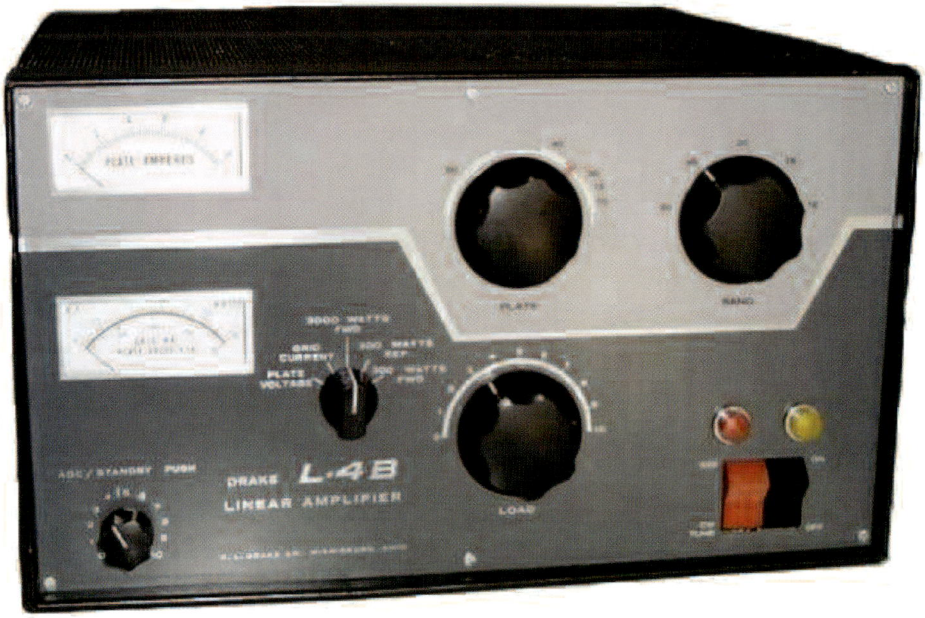

All Drake amplifiers have been built using the Eimac 3-500Z, the L4 and L7 used a pair and the L75 used a single 3-500Z with a built-in power supply. Just like Collins S- line, Drake had the Drake twins T4X and the R4C and any of the three amplifiers matched the line up to give a good looking, quality station. All the Drake transceivers gave around 150 watts including the last transceiver the TR7. Because of this the EHT did not have to be too high to achieve 1000 watts and the amplifiers ran at around 2400 volts compared with the TL922 and SB220 which use up to 3200 volts.

The Drake L7 is the last of the RL Drake amplifiers, using 2 x 3-500Z to deliver 1000 watts and using a power supply that sits on the floor. If you find one in good condition it would be a very nice addition to your station. It is a good reliable amplifier and easy to service.

The Drake L75

The Drake L 75 using a single 3-500Z and internal power supply unit

Unlike the L4 and L7 which only covered 10 - 80m, the L75 also includcd 160m.

There were not many Drake L75 made, compared with the very popular Drake L4 and Drake L7. The single 3-500Z would deliver around 700 watts.

This is the only Drake amplifier to be fitted with an internal power supply. The L75 is well laid out and easy to maintain.

Emtron

A division of Emona Electronics
Email Emtron@ozemail.com.au
web www.emtron.com.au
92- 94 Wentworth Road, Sydney, NSW, Australia

Emtron in Australia, owned by Rudi Breznik, was the first company I came across using the GU74b in the DX1 and DX2 amplifiers. They have been shown and sold at Friedrichshafen for more than 20 years. Now a lot of European manufacturers use the Russian tetrodes which include Acom, Alpin, and OM etc. Over the years Rudi and I have become good friends. When I had a problem sourcing special RF components Rudi helped me with switches and special capacitors before all the components became much more available at Friedrichshaven and on the web. Though we were in competition, our amplifiers were very different as I had never used tetrodes in any quantity and Rudi had never made VHF and UHF amplifiers. I think we both respected each other in the different ways that we made our products.

Above is the DX1 which is one of Rudi's first amplifiers using a single GU74/4CX800A well before any other company started to make Russian tetrode amplifiers.

Emtron DX1 inside

You will see from the picture above the very nice layout of the tetrode amplifier. Unusually the GU74 is mounted on its side which is not a problem as these ceramic tetrodes are made to operate from any position. Later the front panel changed and the meters were changed for a modern LED bar graph which shows all the parameters at the same time, as the next picture shows.

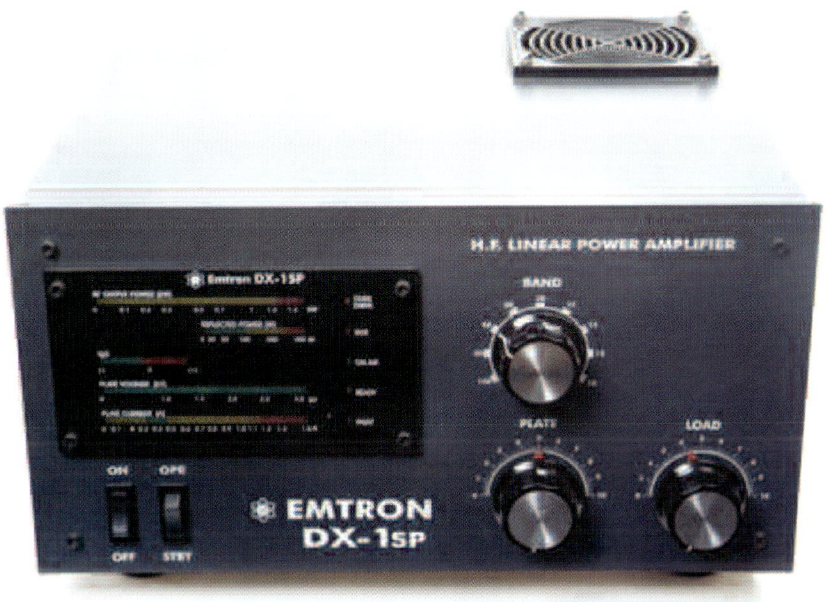

This later Dx1 is part of a family which includes the DX1, DX2 and DX4 all using the Russian ceramic tetrodes.

Heathkit Company

Heathkit was founded by Edward Baynard Heath at its base in Benton Harbour Michigan, in the early 1900s as an aircraft company making light aircraft and parts. Unfortunately he died while doing a test flight in one of his aeroplanes in 1931. In 1935 Howard Anthony purchased the then bankrupt Heath Company and focussed on selling accessories for light aeroplanes. Howard Anthony died in 1954 and the company was bought by Daystrom Company who started to make receivers, FM tuners, audio amplifiers and a vast range of test equipment including a kit oscilloscope for only $50. Daystrom, under the name of Heathkit went into making AM and CW transmitters. Most older radio hams will remember the famous DX100, first made in the 1950s and upgraded to the DX100B in 1959. The smaller DX40 and the phasing adaptor, the SB10U, to get on to SSB, followed a series of AM transmitters, which mainly just sold in the USA. Collins and Drake were Heathkits main competition but Heathkit was very much the everyman's Radio Company

I admired Heathkit with their ready to build kits. Many years later I did the same with my Ranger 811K which sold as a kit, compared with the Ranger 811H sold as the fully working 800 watt amplifier.

Heathkit had excellent handbooks, even describing how long to cut each individual piece of wire between components. The detail was excellent and even if you did not know anything about electronics you could always build a Heathkit and learn while you were doing it.

There is much more history about Heathkit on the Web. Do look it up.

Heathkit SB220

10 to 80m 1000 watts from 2 x 3-500Z

The Classic Heathkit was a real individual of amplifiers. To start with its colour is unique and reflects Heathkit style and quality. In the late 60s and early 70s before ceramic triodes became available the SB220, with its pair of 3-500Z, was one of the highest power amplifiers around. Thousands were sold all over the world and delivered a big punch of a signal. In the UK I remember a very old-timer who was always a big signal on back scatter on 20m and some 150 miles away. Santé G2PU had a homemade Quad at about 70 feet and used the classic SB220. Golf 2 Pacific United, Cambridge, England, as he would say, just dominated the band in 70s and 80s with his SB220. He also had a spare just in case it went wrong! The SB 220 is nicely laid out using conventional chassis with the high voltage stuff in the top and input and control parts on the underside. In many ways it is the Ford car of amateur amplifiers. It functions well but does not have the class of the Drake or the Kenwood TL922. I have repaired many SB220 and the first thing you come to is the fight with the case. The nicely laid out RF box is inside the green case which is a solid wraparound compartment. You have to remove its feet then try and slide the whole thing out, which always seems to get the better of you! You can then remove the aluminium inside covers which are just fastened with self tappers. When you get there everything is easy to get to and nicely laid out. If I have any complaints about the SB220 it would be the components are on the light side for a 1300 watt amplifier. The ceramic switch and the tune and load capacitors all could be larger. They are prone to flash over if slightly mis-tuned. The transmit/receive relay is an open type relay and often suffers with bad contacts. Fortunately it is easy to take to bits and clean.

I think my big gripe about the SB220 would be the helicopter fan on the back panel. It really sounds as if it wants to take off. Yes it does give good cooling but kills all other noise in the shack. It can be slowed down with a series mains capacitor, or you can change it for a good quality fan such as a Papst which aims the air much better at the centre of the anodes of the 3-500Z and is many decibels quieter. Most large amplifiers fit slow motion drives to make tuning easier on the higher bands and this is another aspect that could be improved. If I was renovating an SB220 I would also add a soft start circuit to prolong the life of the 3-500Z. After saying all this, if you find one in good condition at a fair price it would make you a nice 1 kW amplifier.

Heathkit SB200

10m to 80m

The Heathkit SB200 was the first amplifier I bought before I decided to build them, the smaller brother of the SB220. It is well laid out using a pair of 572b which were made by RCA or Cetron. The amplifier would deliver 700 to 800 watts CW and 1200watts PEP. I remember paying £200 for this secondhand amplifier in 1971 at the Leicester Amateur Radio Show, a lot of money, but I thought very good value for money at the time.

A lot of the same components are used in both the SB220 and SB200. I think the SB200 is much more conservatively rated. Flash-over is less likely and always tunes very nicely. I have never had to change a mains transformer in either the SB200 or SB220 so I would say both are well rated. Electrolytics age and do go so if one does go, change the whole group because if one has gone it has probably put stress on the remaining good ones. The outer case is also a solid wrap around like the SB220, access to the inside is by taking off the bottom feet and sliding the inside out. It can be difficult but not as difficult as the SB220!

The SB200 inside is nicely laid out and access to all the major components is quite easy. The TX/RX is often a cause of problems but fortunately is an open case type. You can remove the small fixing spring and remove the poles of the relay to service and clean. Make sure they are all in line before you put it back. The fans in the SB200 are a nice quiet type, though after a few years operation they are worth inspecting as they can seize. Clean first then add light oil or WD40 to the rotating armature. Changing the 572b is quite straight forward. Make sure you put the two large heater pins in the correct holes as it is possible to put the valves in the wrong way round. There is no special setting up with new 572b. Unfortunately all 572bs are now made in China and I do think their long term reliability is not as good as the original Cetron ones. Heathkit made a mobile version of the SB200 called the HA14. The HA14 was basically the SB200 in two separate boxes, RF deck and power supply unit. The unit could be bought with mains PSU or a 12v DC PSU intended for mobile use.

The SB200 is a nice compact amplifier and if you find one at a reasonable price it would make a good medium power amplifier

Heathkit SB1000 10m to 160m

A more modern version of the older amps

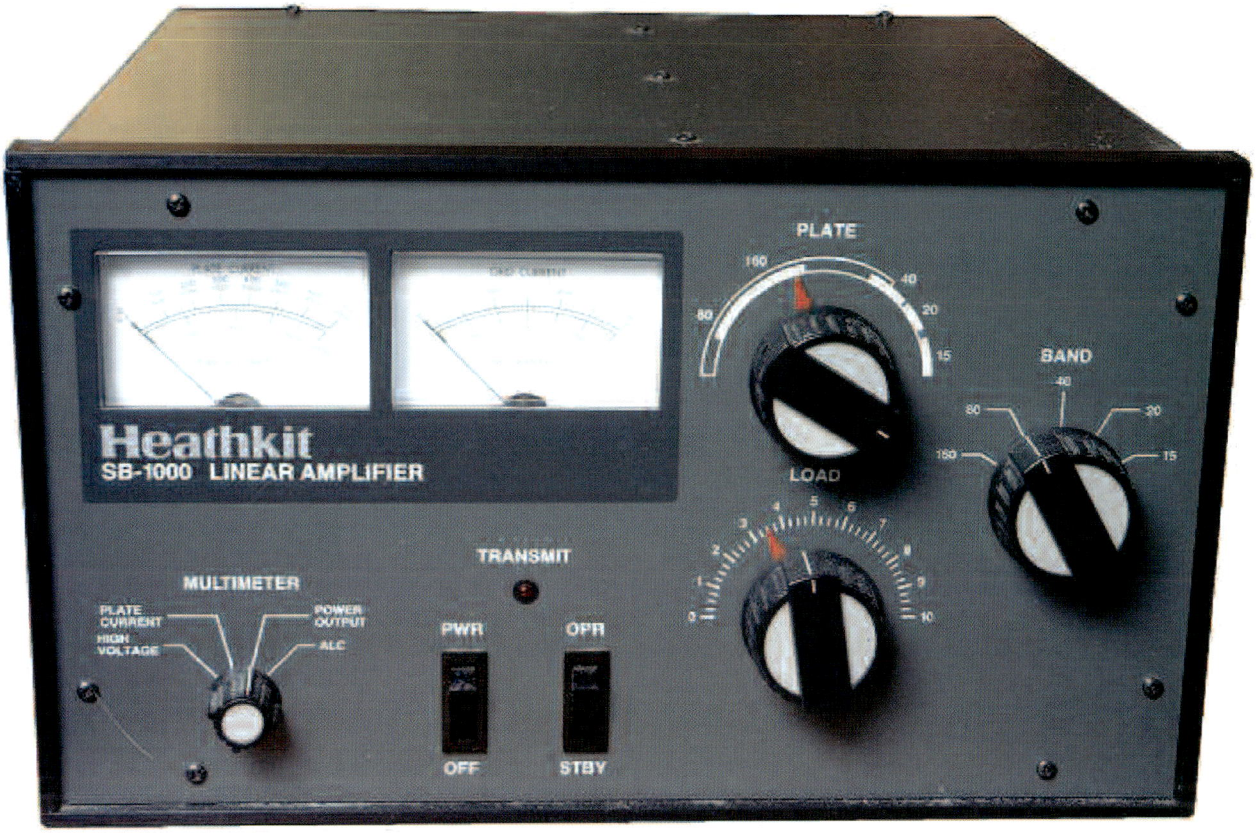

The Heathkit SB1000 and HL 2200 were the first Heath amplifiers to include 160m. This was the company's last attempt to revamp the amplifier business although the SB1000 and the HL2200 were never as popular as the original famous green SB200 and SB220 amplifiers.

Valve amplifier manufacturers started to move away from the glass valves with the availability of the new ceramic triodes and tetrodes. The case was much improved for access than the older amps. Slow motion drives were also fitted and the metering improved. An operate / standby switch is also fitted which was never fitted on the earlier SB200 or SB220. An easy item to fit in series with the TX relay PTT.

The single 3-500Z delivers 700+ watts from what I think is a nice compact amplifier.

The HL2200 is an improved version of the SB220 delivering 1300 watts CW 2000 watts, PEP, 10 to 160m.

Heathkit SB230

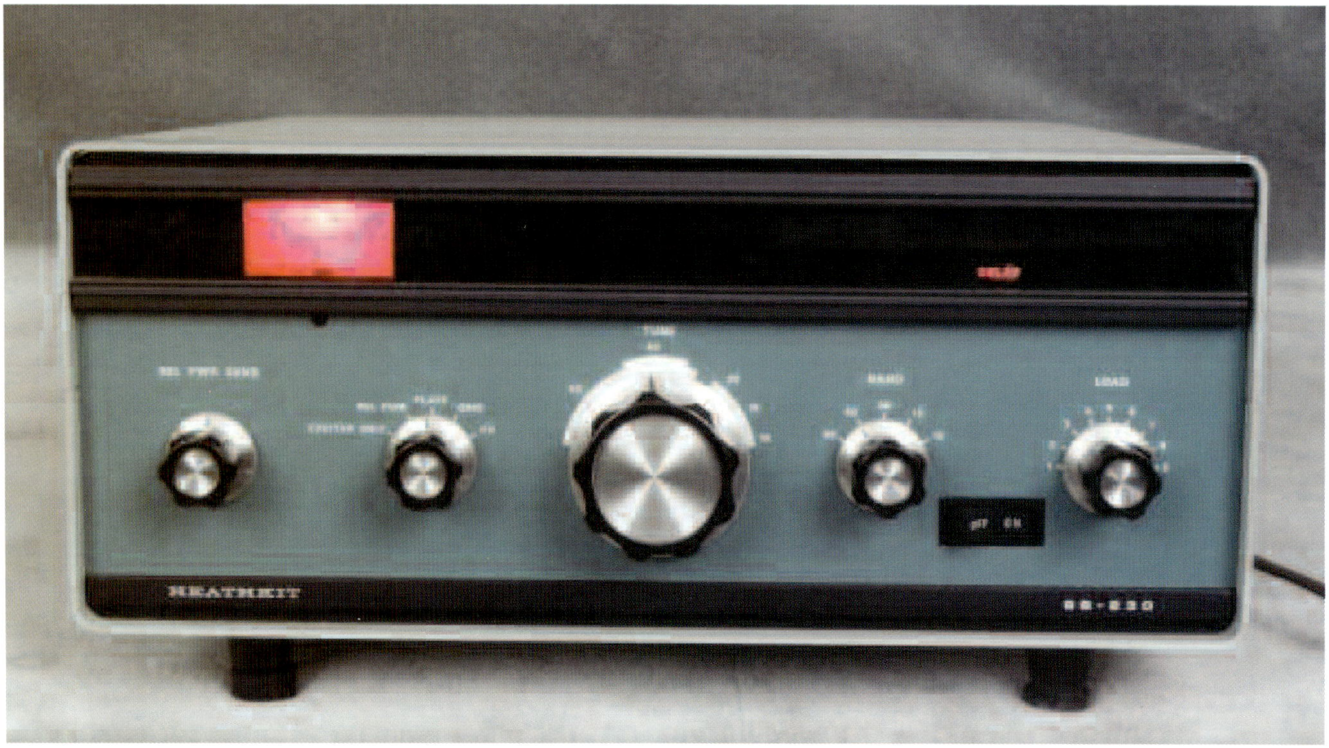

The Heathkit SB230 was the only amplifier that Heathkit made using a ceramic triode. It uses a single Eimac 8873 in a grounded grid configuration and, I believe, this was the first amplifier to use one of the newly developed ceramic triodes from Eimac. It never sold that well but was made to go with the modern SB104 which had been brought out at the same time. This is the only amplifier I have never had to repair and in the UK is very rare. The only one I have seen was owned by ZS1FP in Cape Town, South Africa during my time at sea. For some reason 160m was not included.

Linear Amp UK Explorer

The Explorer was the first commercial amplifier I designed similar to the Heathkit SB220 and the Kenwood TL922 employing a pair of Eimac 3-500Z. Earlier models did also use 4-400 tetrodes in grounded grid. Early models only covered 10 to 80m, but later ones like the one shown below did 10 to 160m including the WARC bands. From the start all my amplifiers were fitted with soft start as standard to eliminate the stress on the heaters.

In the section about Linear Amp UK, I will show a lot more detail and include circuits and fault finding on 3-500z Amplifiers.

Explorer 1200 top inside

This early model Explorer shows the two iron core transformers, one for the HT and the other for the valve filament. The cabinet is zinc plated steel. The RF deck and power supplies, control circuits are all in separate compartments for easy access to get to all the major components. Later Explorers used a large toroidal transformer which combined the HT, heater and 12 volt control windings.

Linear Amp UK Challenger I

The Challenger was named like the Explorer, to get you out there and cross new frontiers. Like the Ten Tec Titan and Alpha amplifiers it uses the modern 3CX800A7 ceramic triodes. It employs soft start, 3 min start up timer and grid protection. The valves are indirectly heated so you must wait 3 minutes for the cathode to be fully heated. The Challenger was the first big amplifier to use a high voltage toroidal transformer which made it lighter for shipping though still weighed in at just under 30Kg. The demand for the Challenger was great especially amongst the low band DXers

The Challenger has developed with various changes and improvements. It uses either the 3CX800A7, 3CX1500A7 or the Russian GS35b.

The Challenger IV is still made by Roger Banks at the DXshop in Wales, do look it up online.

OM Power Amplifiers

930 30 Bac 126
Slovakia
www.om-power.com

From seeing OM about 6 yrs ago I have been watching with interest their development of new amplifiers. OM has come on from strength to strength making high quality power amplifiers. All their designs are based on Russian tetrodes and if I was making amplifiers now I am sure I would also be using a similar Tetrode design.

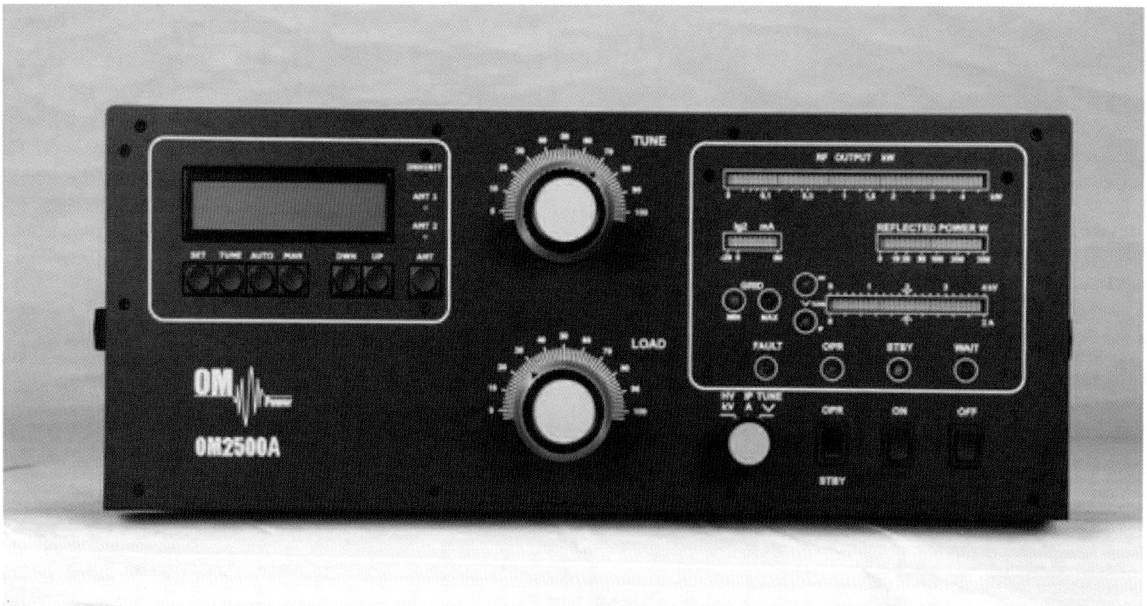

OM 2500A

This is the OM 2500A and when you look inside the amplifier you do get a feel of quality. The valves are reasonably priced if one needs replacing but like most indirectly heated ceramic tetrodes, if run properly, the valve should give you thousands of hours use without needing to be changed. All the amplifiers are very well protected in every way and I would recommend one after seeing their development over the years at Friedrichshaven. Due to the large output power of up to 2500 watts the input will be around 5000 watts off a normal 230 VAC supply. This means the current taken would be around 20 amps. Their answer, unusually, is to fit 2 power cables to split the load. As I did with some earlier Challengers it uses 2 x toroidal transformers in parallel to supply the power for the rectifier board.

The GU84b Tetrode used in the OM 2500

Above the RF deck from the side

AC input   AC input

OM 2500A Specification

1.8- 29.7 Inc WARC   2500 watts all modes   Drive 40 – 60 watts

As you can see from the OM 2500 block diagram it is a grid driven tetrode amplifier with normal anode, screen and bias supplies. It is like most modern designs where the transceiver drives a 50 ohm load and a portion of it is fed into grid 1. This makes for a very stable amplifier. Unusually you will see the AC1 and AC2 as the 2 mains input supplies. You will basically need 2 x 16 amp wall supplies to run the amplifier at full power.

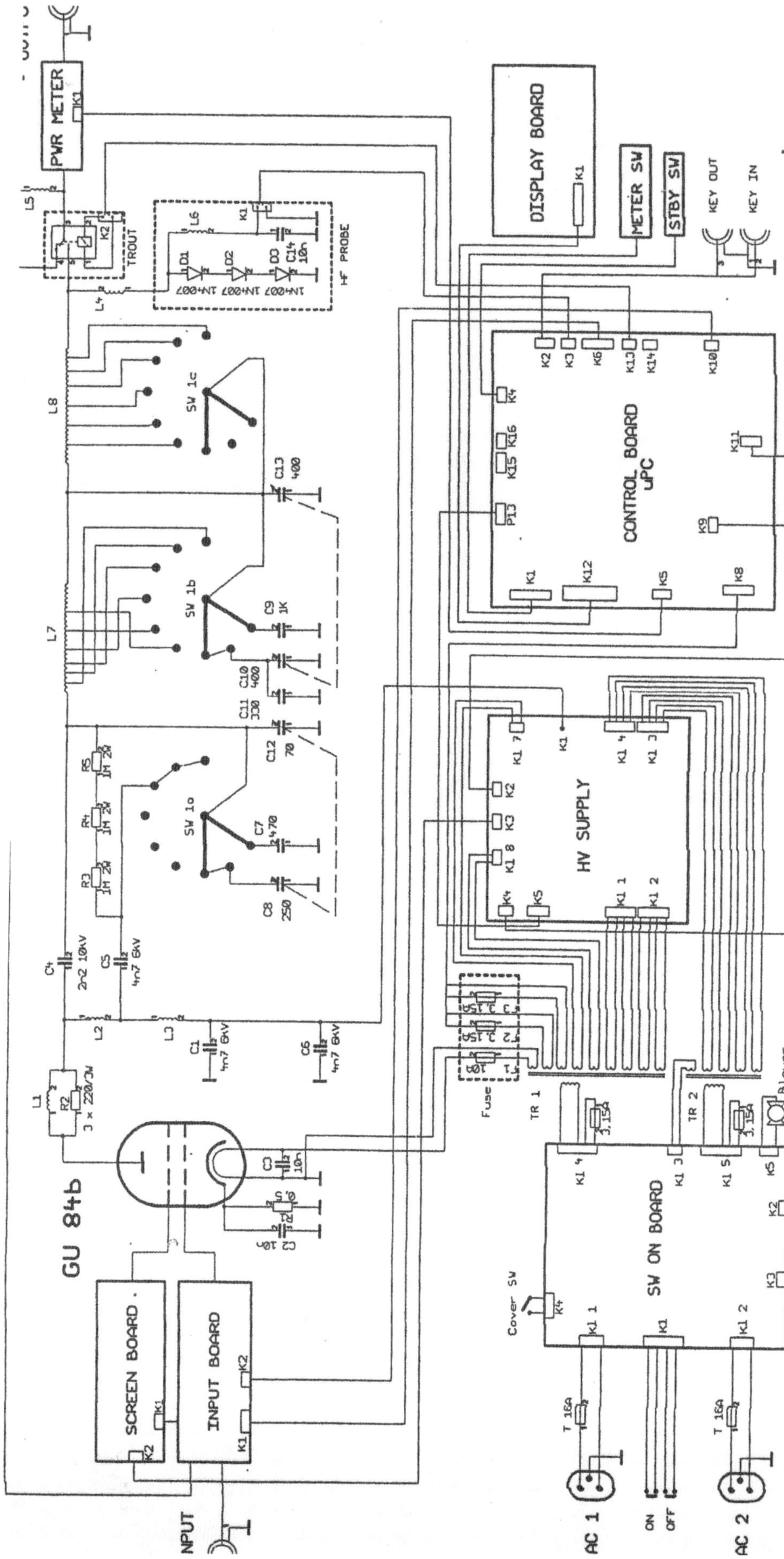
GU 84b
PWR METER
K1
L5
TROUT
K2
L4
L6
D1
D2
D3
C14 10n
1N4007 1N4007 1N4007
HF PROBE
DISPLAY BOARD
METER SW
STBY SW
KEY OUT
KEY IN
L8
SW 1c
C13 400
L7
SW 1b
C9 1K
C10 400
C11 330
C12 70
SW 1a
C7 470
C8 250
R5 1M 2W
R4 1M 2W
R3 1M 2W
C4 2n2 10kV
C5 4n7 6kV
L2
L3
C1 4n7 6kV
C6 4n7 6kV
L1
R2 3 x 220/3W
C3 10n
C2 10n
R1 5
SCREEN BOARD
INPUT BOARD
K1
K2
INPUT
CONTROL BOARD
uPC
K2
K3
K6
K13
K14
K10
K4
K16
K15
K11
P13
K9
K1
K12
K5
K8
HV SUPPLY
K1 7
K1
K1 4
K1 3
K2
K3
K1 8
K4
K5
K1 1
K1 2
Fuse
10
TR 1
TR 2
3,15A
Blower
SW ON BOARD
K1 4
K1 3
K1 5
K5
K2
K3
K4
Cover SW
K1 1
K1
K1 2
T 16A
T 16A
AC 1
ON
OFF
AC 2

Ten - Tec Titan 425

The above picture shows the front Panel of the Ten-Tec Titan

The Titan has simple clean lines and easy to use controls with a low profile.

After repairing a Titan I was impressed with the amplifier design and also the first time I had come across the Eimac 3CX800A7. It excelled anything I had seen in the past for an amplifier delivering 1500 watts from a pair small of valves. Running at around 2400 volts is quite low HT for a 1500 watt amplifier. Ten-Tec like myself moved away from the 3CX800A7 when the price went up and has chosen to use the Russian GU74 tetrode which is a similar size valve but is a tetrode.

Specification

10 to 160m CW/SSB

1500 watts output with 60 watts of drive.

Power supply for HT is in a separate box which can sit on the floor

Ten-Tec also makes the Centurion amplifier using 2 x 3-500Z.

Ten -Tec is now part of the RF Concepts Company.

Below the inside of a Ten Tec TITAN

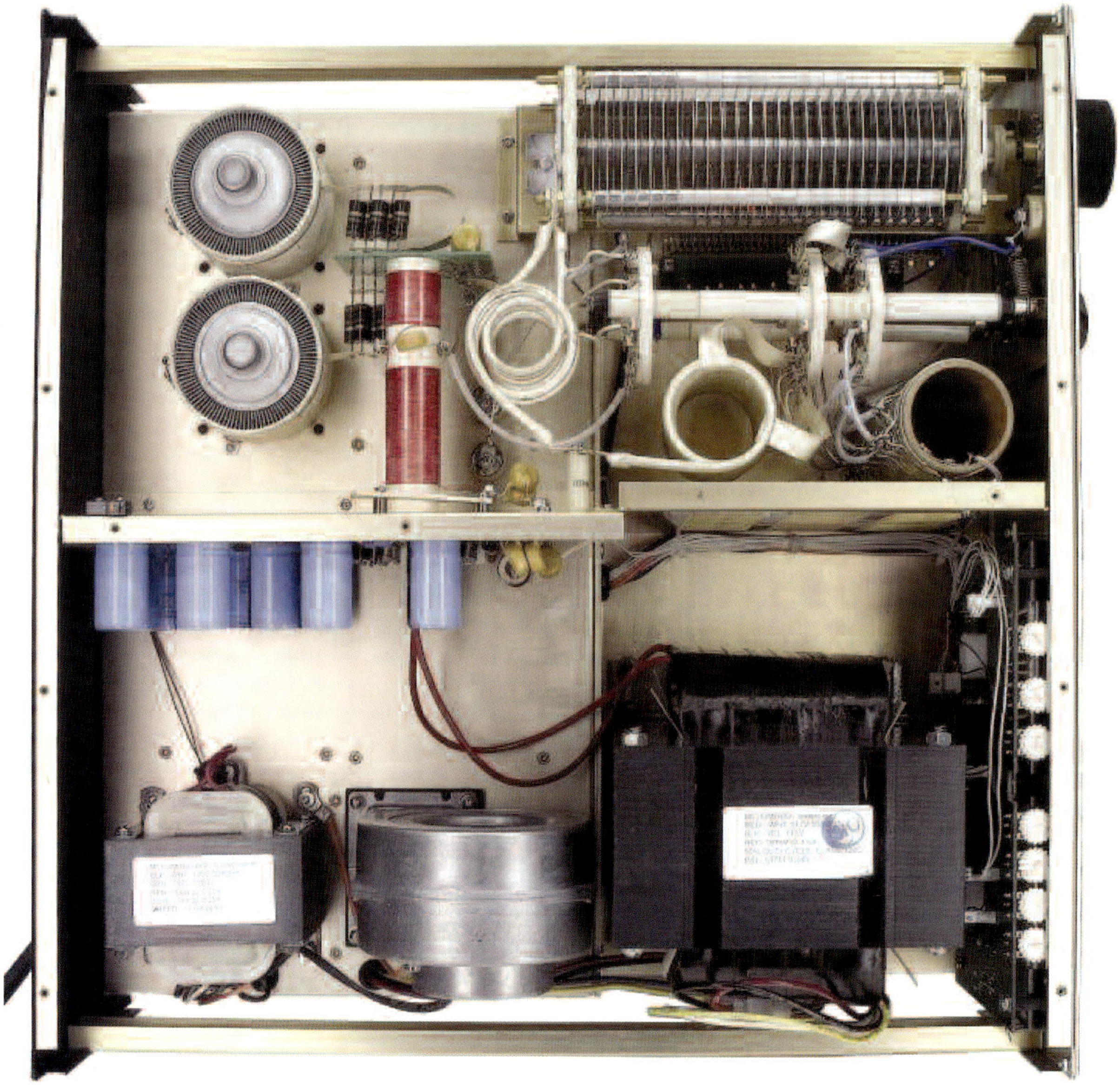

Very well laid out and all the components are easy to get at which is nice to see. The HT transformer is mounted in a box which sits on the floor. This gives the Titan a very low profile. I would recommend anyone to buy one if you came across one in good condition at the right price.

This particular version is using 2 x GU74 but the majority use the well proven 3CX800A7. Of course the 3CX800A7 is a high µ triode and the GU74 or 4CX800 is a tetrode.

Yaesu Musen Company.

Yaesu Musen must have made more amateur radio amplifiers than any another company in the world. The Company was started by Sako Hasegawa JA1MP in 1959 near to Tokyo Japan. Sako died in 1990 and the company was taken over by the managing Director Jun Hasegawa. In 1998 the company acquired the Standard radio company from Marantz and they renamed the Company to Vertex Standard Co Ltd.

The first Yaesu FL1000 linear amplifier

10 to 80m and like others at the time used 4 x glass TV line time base tubes

It gave around 500 watts, a lot of power for the 1960s

Designed to go with the FR101 and FL101, later the FTDX150 and FT200.

The Yaesu Musen FL2000B

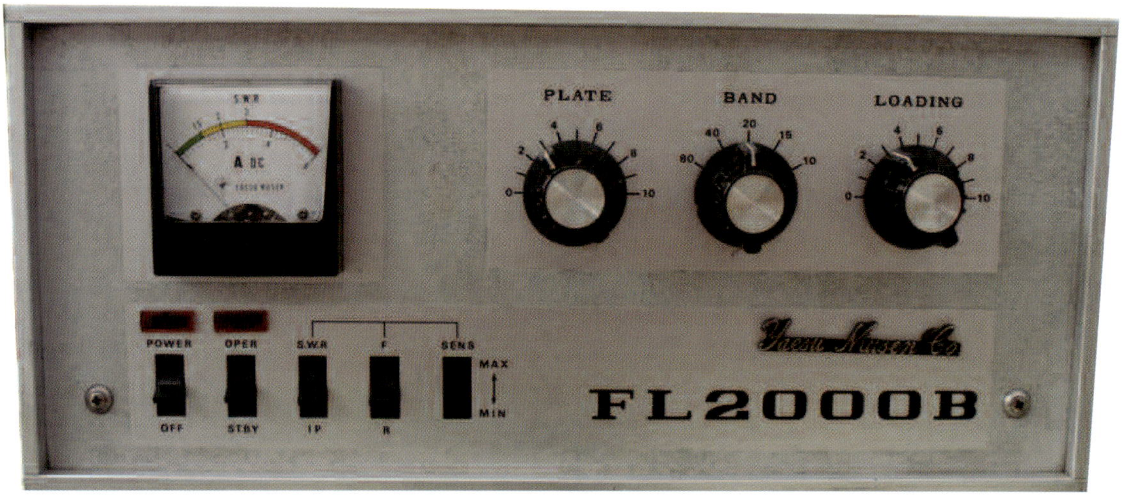

The FL1000 and FL2000 came out in the late 60s and early 1970 to match their range of transmitters and transceivers. The early amplifiers used the commonly available television line time base valves which were very cheap and cheerful at the time. The later FLDX2000 (B) came out with 572B triodes and later renamed the FL2100. Yaesu have always made matching amplifiers to go with their range of transceivers. All the valve amplifiers continued with name FL2100 number but have had many changes to improve it over the years - the FL2100, FL2100B and finally the FL2100Z. The Z had two models, early ones did not have the WARC bands but later ones had separate switch positions for all the WARC bands. The FL2100Z has extra relays added for aerial switching plus one used now for a low voltage PTT switching circuit to suit some modern radios that can only sink 12 volts at 100mA or less.

The Yaesu FL2100Z

10 to 160m 700 watts

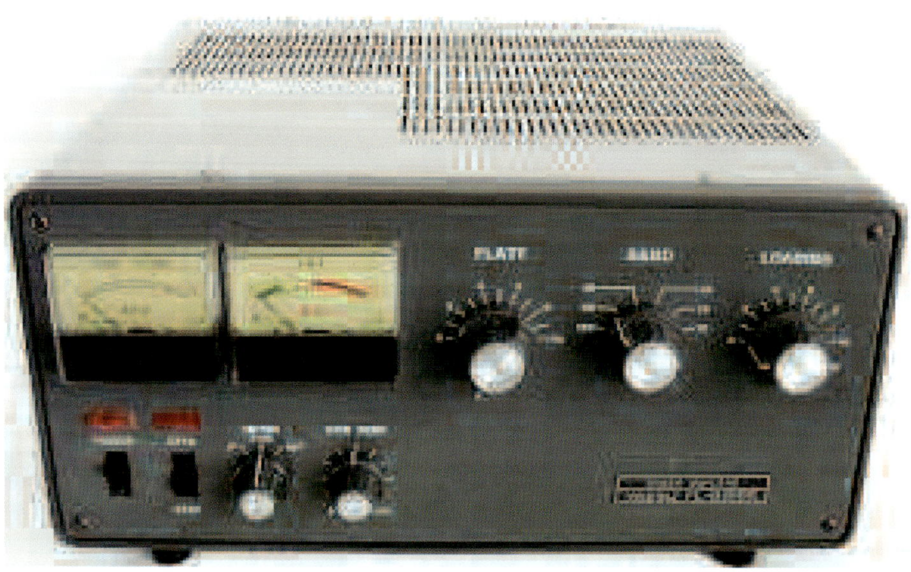

The Yaesu FL7000

After the long term run of the FL2000 and the full FL2100 series it was time to go solid state with the FL7000. A very nice 500 watt amplifier and due to the time it was brought out looking more like a 80s video recorder. At the time I do not think most amateurs were ready for the change to high power with transistors. The FL7000 has proven to be a very reliable amplifier with its built in aerial tuning unit. It matches any of the modern Yaesu modern transceivers.

The Yaesu Quadra 1000 watt 10 to 160m solid state linear amplifier

The Quadra fully automatic amplifier uses the popular MRF 150 power transistors just like the Icom PW1 and the Expert amplifiers running at 50 VDC. It follows the radio on each band and selects the last aerial used on that particular band. The power supply unit is made to go on the floor, out of the way. The amplifier also includes a fast memory aerial tuning unit.

Chapter 6

The Kenwood
Including Specification
Description
Fault Finding
Circuit

The Kenwood TL 922

Specification

Transmit Frequency Range	160m through 10m Amateur Bands
Excitation	80 watts to 120 watts maximum
Mode and Duty	SSB continuous for 30 min, CW, RTTY continuous for 10 min
Rated Anode input	2 kW PEP, 1kW CW and RTTY
Typical RF out CW	800 watt in the CW position
Typical RF out SSB	1300 watts in the SSB position
Plate Voltage	3.2 kV SSB, 2.2kV CW/RTTY
Circuit Type	below -30dB
Input Impedance	50 to 70 Ohms unbalanced
Output Impedance	50 to 70 Ohms unbalanced
Cooling	Axial fan
Interlock safety to earth	High voltage cut-off when lid is removed.
Thermal Protection	Locks out transmit if HT transformer overheats.
Fan motor delay stop time	140 seconds after mains switch off.
ALC output . (Back panel)	Negative going, adjustable threshold -8 VDC max
Power Tubes Used	Eimac 3-500Z, x 2
Semiconductors	Diodes 18, Zener diode 1
Power requirements	120v @28 amps, 220-230v @14 amps, 50/50 Hz
Size	390mm x 190mm x 407mm
Weight	31 kg (68 Lbs)

Top Inside

The inside is well laid out using good quality capacitors and switches. The picture shows a lit up pair of original Eimac 3-500Z. The HT Transformer can be seen in the top left corner.

The Kenwood TL922

The TL922, over the years, has been one of my favourite amplifiers. It was introduced in the early 1970s to go with the TS520/530S and the TS820/830S valve PA transceivers, the first solid state transceivers TS180S, TS430S and then later the TS930S, TS940S, TS950 series Base Transceivers. I was licensed in 1970 and there is no way I could afford a TL922 at the time, so I decided to copy the 922. In the 70s the handout leaflet for the 922 had a circuit on the back. Very useful!

It was very kind of Kenwood! Though I had to keep getting a new leaflet as I was using it in the work shop, it soon became tatty.

After owning and repairing many TL922 I have always been very impressed with the design of the amplifier. Unlike some articles on the web, I believe an untouched TL922 is one of the most stable amplifiers ever built. It does not need any modifications. I, personally, have never found one with any with parasitic oscillation problems.

It has a very well designed front panel compared with many American amplifiers and was one of the first to use a proper injection moulded front panel. Access is easy to the most important parts that might go faulty. The only exceptions are the tuned input circuits and the heater transformer.

I think the TL922 would have been better if it had been fitted with a soft start circuit to minimise inrush current on the heaters and the EHT on switch on. This can easily be added if you feel that way inclined.

The amplifier is well cooled, though cooling to the valve bases could be better. Air circulation to the base of the 3-500Z is important as each valve takes 15 amps at 5 volts, some 75 watts per valve. Eventually the valve pins can become very hot and the pins internally can become a dry joint - a point to check. If the valve filament does go out, with a large 200 watt soldering iron you can de-solder and resolder the centre of the valve pin! This often repairs the valve!

Nothing on the TL922 has been compromised, nice tuning dials with slow motion drives on both the tune and load capacitors. It has a good cooling fan that very rarely goes noisy, if it does, it can be replaced with a Papst fan 4850 and works just as well.

I have never known an HT transformer go faulty on a TL922 but I have known several heater transformers go. This is normally due a short-circuit turn on the transformer and will smell like very hot varnish. If the heater transformer does go faulty, they are still available from Kenwood (at the time of going to press). To change it is quite straight forward, just remove the side of the amplifier using Allen keys and then you can access the transformer.

The TX/RX relay is very substantial, I have never had to change or repair one, and in fact I have never had to even clean the contacts of one yet!

The 3-500Z valves

The original valves that came with the TL922 were Eimac with pressed metal anodes and many are still in use after some 30 -40 years. Under normal operating conditions, at 1 kW output, the valves run at dull red which is quite normal. If the valves are changed to any other make, no setting up is required, the bias is the same, just plug them in (making sure the HT has disappeared before touching the anode clips!). The later valves from China and the Amperex valves made in France are fitted with graphite anodes as indicated by the number on the valve i.e. 3-500 (ZG). These run cooler as the mass of the carbon anode is larger and gives better cooling.

I would not run one graphite valve with a pressed metal type, it is just like putting different makes of tyres on a motor car. The pressed metal anode is supported by its own single pin moulded in the top of the valve. If the valve has had a heavy jolt the anode can bend on its support pin from the top of the valve. This can cause an anode-to-grid flashover. This will also cause the grid choke to go open circuit and in some cases it can take out the meter shunt. The graphite version of the 3-500 ZG is more robust and anode distortion is less likely.

I believe that before Eimac stopped production of the 3-500Z a run was done making the 3-500ZG. It is a pity that they stopped production, as I am sure it would have been a better valve and popular if the price was kept down. The Chinese, in their wisdom, stepped in and started to produce the 3-500ZG. The early ones were a bit un-reliable but now I have full confidence in the Chinese 3-500ZG. I would recommend them to anybody and the price is also attractive, but creeping up.

Power Supply

The TL922 uses two transformers, T1 is the HT transformer part number LO1-8026-05 and it can be switched from medium to high voltage (2200 or 3200 volts DC). It is marked on the front panel as CW or SSB. This is done on the transformer by switching between 800 volts AC and 1200 volts AC. A standard type of voltage doubler circuit is used to deliver the HT voltage (AC x 2.84). If you do not want to use more than 800 watts it is worth keeping the voltage selector to the CW position (2200 volts), this is fine on both CW and SSB and there is then less chance of the valve suffering internal arcing. If you want to use more than a kW on SSB, switch to the SSB position. This gives you an RMS power of around 1300 watts which is around 2000 watts PEP.

Smoothing is done with 8 x 200µF 500V capacitors, mounted in a special housing in the side of the amplifier, decoupled with .01µF capacitors and discharged with 47K bleed resistors.

Heater Transformer

T2 is the Heater Transformer, part number LO1-8038-05, the winding is a 10V winding, centre tapped to give 5-0-5 volts. Each valve is connected to one side only 5-0 or 0-5. The cathode return circuit is connected to the centre tap of the transformer and pulls the Heater /Cathode circuit down to 7.5 volts in the transmit mode. T2 also has an 80V winding for the relay control circuit and the bias circuit to give a 100V DC supply.

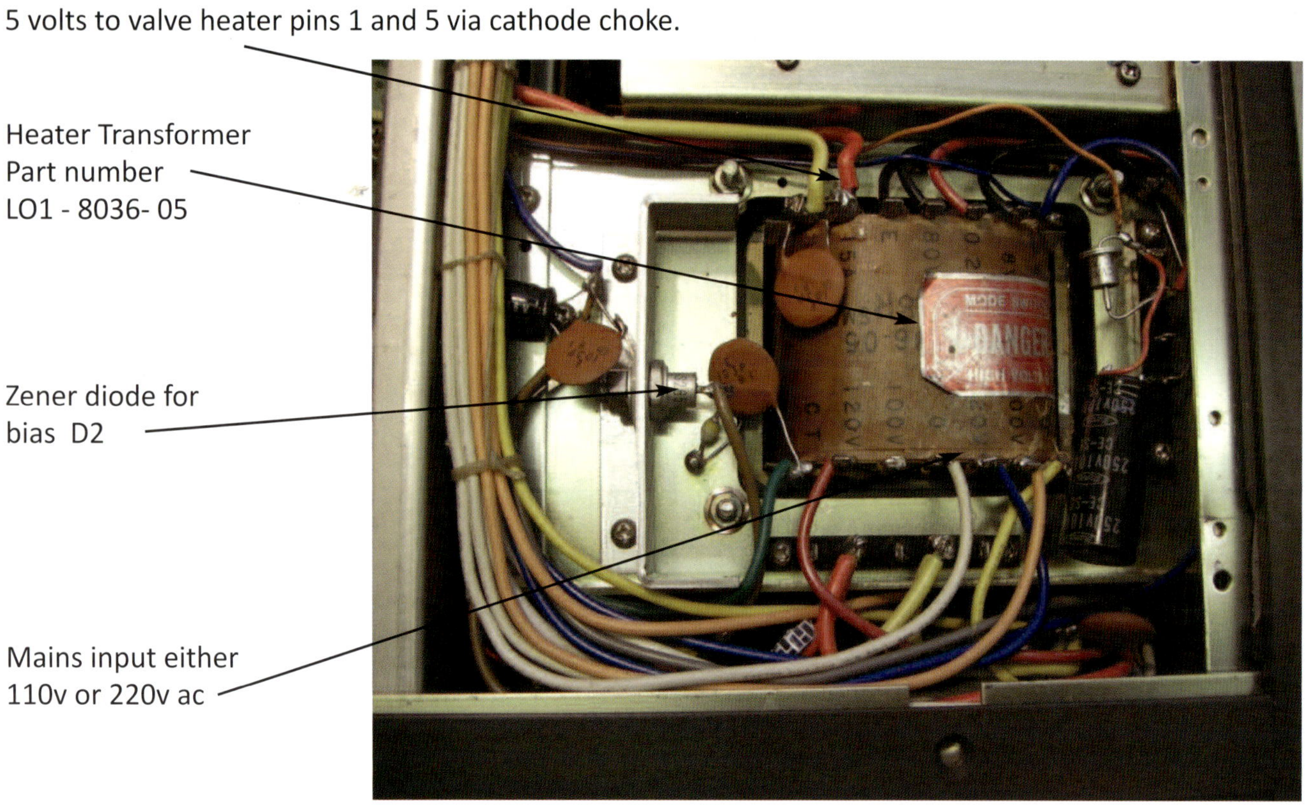

Heater transformer connections

The Heater transformer can be wired on the primary in series for 230v for Europe and in parallel for American 110 VAC supply.

The zener diode D2, is for cathode bias, it is a 7.5V 1S265 but can be replaced with either a BZY 91 or 93 available from Farnell Electronics. The heater transformer is a component that can fail on the TL922, I have changed several and are still available from Kenwood spares.

KI would highly recommend adding a soft start circuit to the amplifier. When you normally switch on the inrush from the mains supply can reach well over 25 amps for a moment, often tripping modern household circuit breakers. There are 2 reasons for this:-

1. Charging the electrolytic capacitors to full DC voltage either of 2200 or 3200 volts, depending on the mode setting, all this happens within the first cycle of the AC mains applied.

2. The resistance of each 3-500Z heater is much lower when cold, when warm the normal resistance is found and each valve takes its normal current of 15 amps at 5 volts, some 75 watts per valve.

A soft start circuit can be installed inside the amplifier. This can be fitted in the fan housing. You can break the incoming mains feed from the on/off switch. The soft start circuit comprises of only 5 components and is very easy to make:

A 24V relay single pole, 2200µF 40V capacitor, 50 ohm resistor at 50 watts or more, a 2K2 resistor at 25 watts and a diode like a 1N5408 (but any power diode will do).

See adding soft start later in the book.

The Kenwood PA coil and the input tuned circuits

The underside of the TL922 showing the PA coil with taps and the extra capacitance for Tune and Load capacitors used on 160m

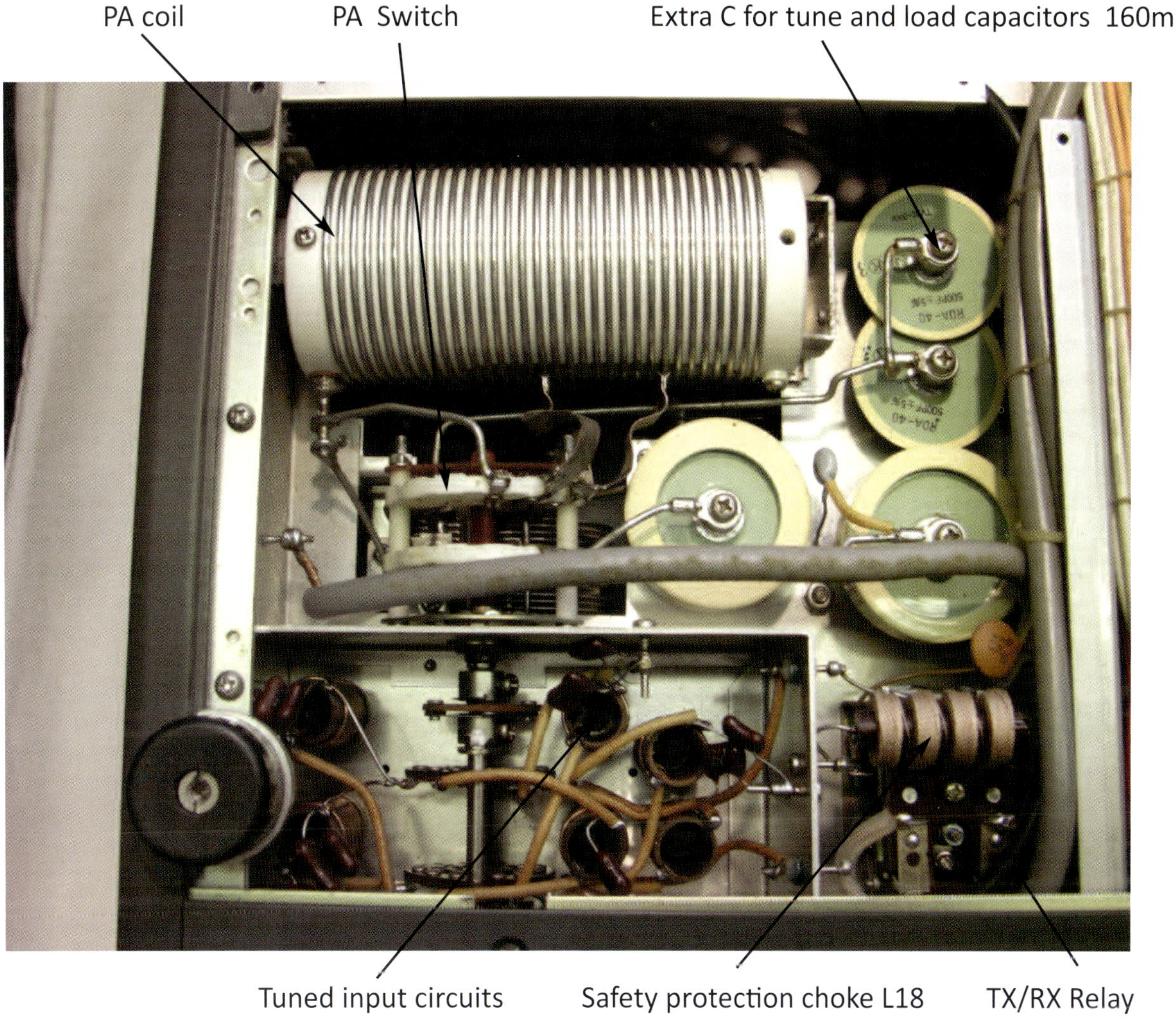

The tuned input circuits that match the 50 Ω input to the cathodes of the 2 x 3-500 valves are not easy to adjust but very rarely need any adjustment.

The TX/RX relay is one of the worst components to get at in the TL922 but luckily hardly ever goes faulty. Protection Choke L18 is the HV protection. In the event that C34, the HT blocking capacitor, ever had the misfortune of going short circuit, L18 would pass all the high voltage to earth and stop any danger of high voltage getting to the aerial.

Front Panel Bulbs

These do make the meters look nice but do go eventually with age. These bulbs are a hard wired type and soldered to posts near to the back of the meters.

The Part number is B30-0048-05 and still available from Kenwood. They come with a special rubber fixing grommet, but you still have to solder them in.

Fault finding in the Kenwood TL922

Low RF output

Often one valve has had an internal flashover and has caused the grid choke L7 or L8 to go open circuit. When choke L7 or L8 goes open circuit, the power drops off and you will see a slightly higher than normal input SWR into your transceiver. They are located on the base of the valve and very easy to change, change with a 3 amp 40µH choke.

The anode to grid flashover shows either the grid or anode has some signs of distortion and could be one valve starting to show signs of failing soon. It is worth keeping an eye on that particular valve.

No Drive

1. When you cannot drive the TL922 it is usually a grid /cathode short on one of the valves. This can be an intermittent due to expansion when hot. The appearance of the valve looking from the top can look perfect and normal (even a relatively new valve can develop this problem, though when a valve has had 3 months of operation or more it will probably last many years). The only cure is to replace the valve.

2. If changing the valve does not cure it, check the TX/RX relay. Go into the TX mode then check for DC continuity between the input SO239 and the input capacitor feeding the cathode of the valves. You should have continuity all the way so this checks the input switch, TX/RX relay and the input tuned circuits. This can be done safely with the top and bottom covers removed. The micro switch on the top panel should not be pressed in, this disables the high voltage supply. It does, however allow all the relays to operate normally so the test can be made. First remove the flapper on the aerial relay and clean the contacts and make sure the contacts are all in line. It is very unlikely that you will experience a fault on the input switch but it is worth checking that the wiper is not dirty or bent.

Drive but no output.

The input circuits are OK but you have no RF coming out of the amplifier.

First check the HT voltage on the front panel, this should be in the range of 1800V on CW and 2400V in the SSB mode. The transformer primary goes via the top panel leaf switch to disable the HT when the top cover is removed. You can see if the HT voltage is alright on the front panel volt meter.

Check the RF compartment for any sign of burns around the PA Pi network coils and the large ceramic switch in the Pi network.

Check the anode blocking capacitor for being broken or open circuit, very unusual but possible. Part number C34 C91-0408-05, 2000pF @ 15kV. If this goes short circuit it will blow one of the mains fuses as it will sink the high voltage to earth through the protection choke fitted across the load capacitor.

Only one valve lights up

This could be due to a bad connection on one of the valve bases, very common!

Remove the valve and clean its pins and the valve socket. A single valve can also have a dry joint within the heater pin itself. You can remove the solder with a large soldering iron and replace with new solder. Unleaded solder has a higher melting temperature so it is better for the high temperature that the pin runs at. It is worth a try as the valve could still be good. You have nothing to lose by trying the repair. You will need a large 200 watt soldering iron, desolder with a solder sucker and resolder with new solder. When the valve is cold, test for continuity - the valve should show a low resistance. When hot the dry joint appears and shows a high resistance and stops the heater from lighting. The heater is running at 5V at 15 amps some 75 watts per valve.

Space is limited and the cooling could be better on the valve pins.

The picture below shows the valve bases on the underside of a TL922

The two grid chokes shown are fastened to the valve grid pins which are pins 2, 3 and 4. They are 40µH. These two chokes are the first thing to go open circuit when a valve flashes over from anode to cathode. They are easy to replace, but they are often the first sign of a valve showing stress and internal break down.

The two pairs of valve pins nearest the centre of the picture are the valve heater pins. These are the ones that suffer from overheating, and from time to time worth cleaning if you have a problem with a valve heater not lighting.

Bad input SWR on one band only.

The input tuned circuits are individual to each band and if a capacitor goes either open circuit or short circuit it will give you a high SWR reading on one particular band. If short circuit you will not be able to drive the amplifier, all the drive will go to earth. If open circuit you will still be able to drive the amplifier but with a higher SWR reading than normal. This can be temporarily resolved by using the internal antenna tuner in your transceiver to correct the bad match to give you almost full drive again. This is only a temporary cure and you should find which capacitor has failed and change it. It will be a ceramic type and all values are available from Farnell Electronics.

High standing current plus a high input SWR

1) One valve has gone short circuit between the grid and cathode. This is due to age or a severe over drive and the grid has collapsed. All the drive goes to earth and the valve loses grid bias and the standing current is higher than normal. Change the valve for a new one

This affects both valves as the valves are connected in parallel. You can run the amplifier with only one valve connected, try it on 40m. The tuning will be slightly different but it will produce 400 W O/P in the CW position. Make sure the anode top cap is clear of anything before you switch on. This is the easiest way to check which valve is faulty. It is always best to change both valves, but due to cost, if the other valve is not very old you can still use it but the newer valve will do more of the work. Eventually the two valves will balance.

2) The cathode choke and the bias choke have decoupling capacitors to earth. If one of these goes short circuit you will get the same symptoms. You will be able to measure a short circuit to earth with or without the valves plugged in. The normal cold resistance of the cathode circuit to earth is over 1k Ohm.

No Anode input Current

The relay that does the cathode switching very rarely goes faulty but it is worth checking. The 7.5 volt zener diode that sets the cathode voltage can go open circuit and will stop current flow. This can happen after a severe anode-to-grid flashover. The zener diode is found under the amplifier and looks like a silver stud. To test it put the amplifier in the CW position and short out the zener diode - if the amplifier comes to life then this proves the zener is faulty. Change the zener - the original is a 1S265 but a BZY91 or BZY93 7.5V works fine.

I have never known a zener diode to go short circuit on a TL922, if it did the amplifier would still work but the standing current on the plate meter would be higher than normal reading.

No transmit relay operation.

1. If you get no relay operation, this can be due to a fault on the cathode circuit of the valve or the valve itself. This is due to the 100V DC supply having dual use, first to operate the TX/RX relay and second to add a 100V DC supply as bias to the cathodes of the 2 x 3-500 valves.

2. The TL922 has a very substantial TX/RX relay and hardly ever needs any maintenance. It can be cleaned, but it is in an awkward position so it is a good job that it very rarely fails or goes faulty but if the coil went open circuit that would stop the relay working.

3. The PTT lead between your transceiver and amplifier could have been pulled and a wire has become disconnected. Check this first!

4. Another cause can be if you have had to reset your transceiver or have bought a new one. The default for most transceivers is relay off. Go into your radio menu and activate the relay for switching of non QSK amplifiers.

NB The TL922 has a 100 V DC switching circuit. If you have a modern transceiver, check that the voltage is OK to sink 100V DC on the standby/PTT line. Most Base station radios like the larger Kenwood, Icom and Yaesu radios have a separate isolating relay which is fine, but small radios like the IC 706 and IC7000 and some of the Yaesu transceivers can only sink up to 12V at a max of 20mA. Do make sure so it is OK to do this job from your dealer. If it will not an interface is available to go between the radio and amplifier.

Heater Transformer Smells

The most common reason for this is when the heater transformer develops a short circuited turn. The transformer does not blow the input fuse as the current does not exceed the 15 amp protection fuse. The current does not rise greatly, like the old Weller soldering iron with a single turn, the element of the transformer does the same thing and the shorted turn becomes red hot and gives off a very pungent hot varnish smell. Transformers are available from Kenwood, part number LO1-8026-05 and you can remove the transformer by taking off the bottom and the left side of the case using Allen keys.

Don't forget to make a drawing of the wires before removing the transformer. The tray holding the transformer can be removed but is a bit tricky. Other smells could come from the power supply of the HT transformer, but this is very rare.

Intermittent Receive

This could be due to the antenna relay having dirty contacts. It could be down to normal use but if you have had a high static crash or you have been close to lightning. The relay would suffer first. Dismantle and take out the moving part of the relay and clean with a very fine abrasive. Try this before actually fitting a new relay, it normally works

Modifications

In my opinion the TL922 is a very well made amplifier. It is not a high gain amplifier like some of the ceramic triode designs or tetrode amplifiers around. With a gain of just over 10dB or x 10 the stability is excellent. I have seen some of the modifications that have been written about the TL922, but none improve the amplifier operation or stability. Maybe adding more fuses to the HT supply or heater transformer might make you happier! But just how far do you go? The APC hardly ever get hot and I think the stop frequency of the APC is just right.

Soft Start

I do think an internal or external soft start circuit is a good idea. It helps limit the switch-on inrush current, rather than going over 25 amps. This current can easily trip modern circuit breakers in modern houses.

Slow PTT switching

If there is one complaint of the TL922 it is that the TX/RX relay is slow to come in when using high speed CW. It will clip the first dit of your call sign - send 'de' first and you're OK. Sorry, I only use semi-break in when using CW so it is fast enough for me.

Buying a TL922

Even now, some 30 years or more later, anyone buying a TL922 in good condition will be buying a good classic amplifier, enjoy it as you would an old classic motor car ! Look after it and it will get you around the world with ease.

SCHEMATIC DIAGRAM

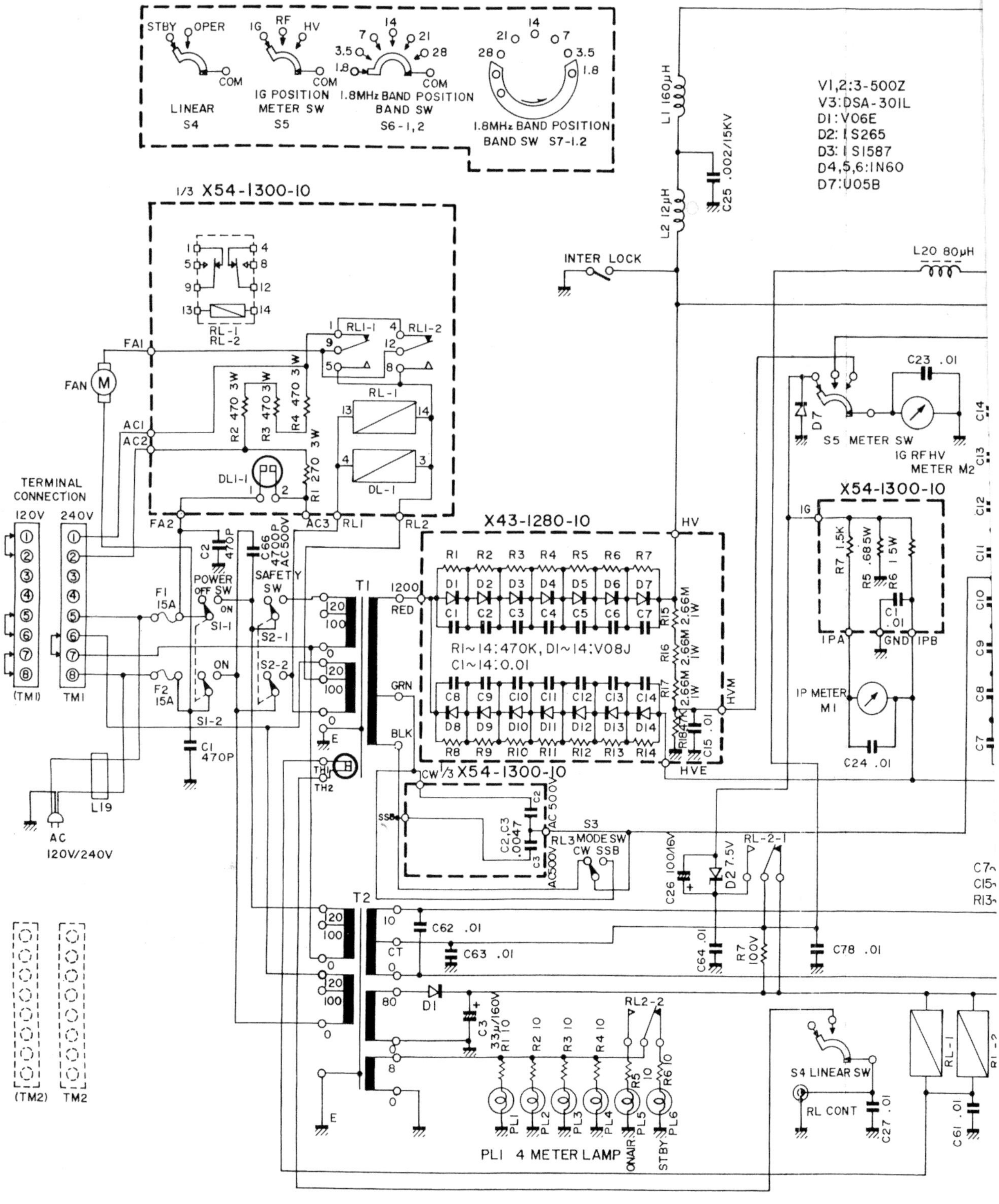

MATIC DIAGRAM

Circuits and specifications are subject to change for improvement.

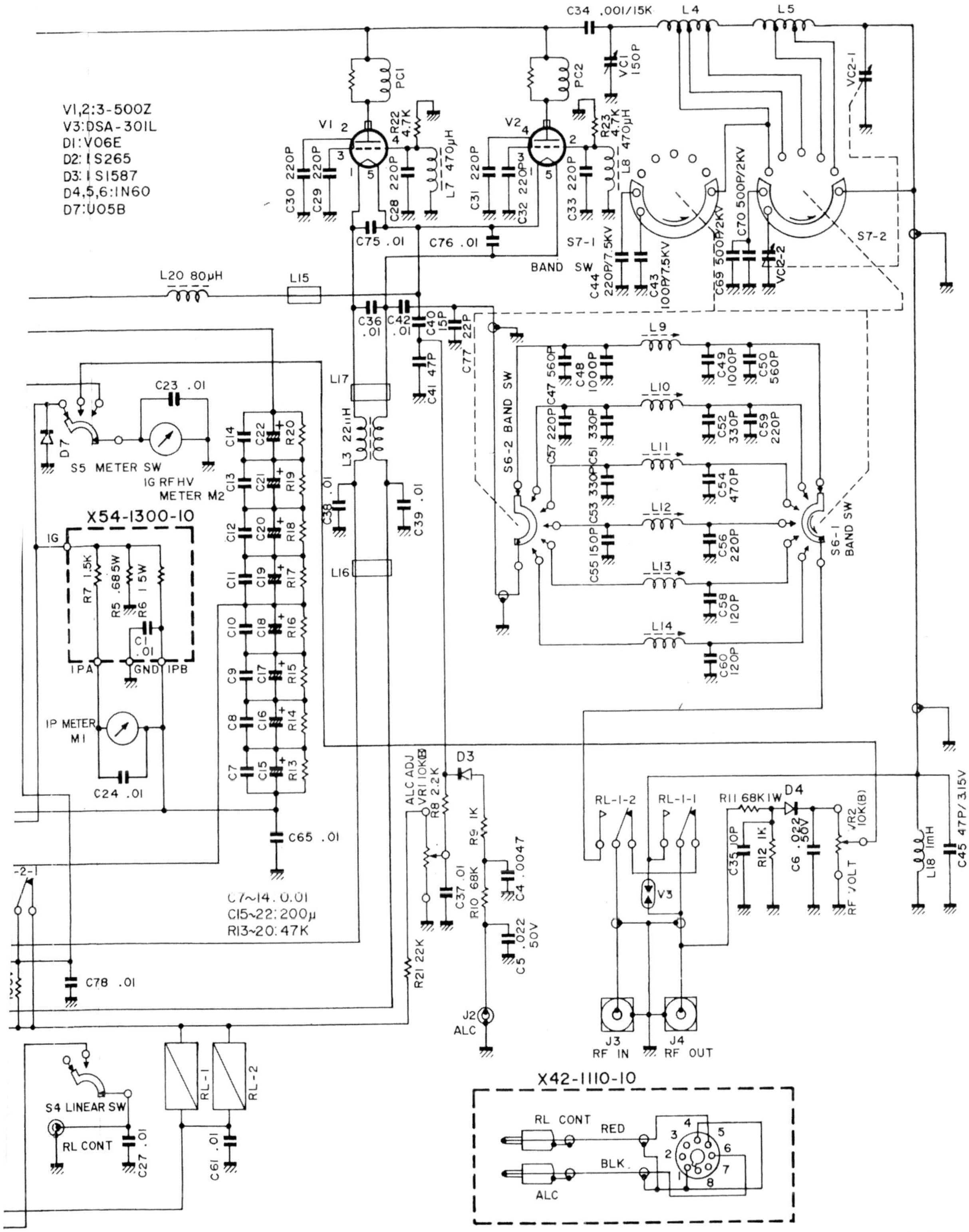

TL-922(K)

Chapter 7

Fault finding on 3-500Z amplifiers

Fault Finding on 3-500Z Amplifiers

DANGER WARNING

Always make sure that the HT voltage has discharged completely and work safe.

Always leave the amplifier for a couple of minutes before touching.

If you do not feel confident touching high voltage components DON'T, it can kill you! Get a person who is a high voltage RF engineer, do not do it yourself unless you are confident and know what you are doing.

If a problem occurs inside 3-500Z amplifiers this section should help you find it. The same fault finding method applies to all 3-500 amplifiers whether they are a single or 2 valve amplifier, this includes the Kenwood TL922, Heathkit SB220, the Explorer 1000/1200 and Ameritron AL80 and AL82. This section should help you find problems

Firstly take a close look inside and then look, listen and smell.

1. Look, due to the high voltages present, components can break down and often you can see a burn or a carbon track from where a flash over has occurred. RF capacitors can split when faulty and you can often see a carbon track or a split in this case. Electrolytics are more often like a fire cracker. Look for them swelling!

Look and see if all the valves are lighting up, if not, one can have an open circuit heater or a valve pin not making a good contact, worth cleaning especially on high current heater valves like 3-500Z.

A 3-500Z can have a dry joint inside the valve pin, it is worth checking, use an ohm meter on the kilo-ohms range, if there is any reading at all try desoldering and resolder pins 1 and 5.

Look at the plate meter, this can tell you so much. A high standing current on the plate meter when the amplifier is not keyed, indicates that there is either a grid/cathode short on one of the valves or the heater transformer has gone short to earth on the secondary. You can normally remove a suspect valve and try the valve that is still in the amplifier. The amplifier will run with one valve removed, you can then swap valves and check the other. Warning, make sure the HT has discharged and earth the anode before touching.

Look at the anode, often when the valve has been too hot the anode will distort and become off centre. The anode is only supported from the top pin that comes from the top of the glass envelope.

2. Smell, a transformer that is failing can still work but be faulty, especially if a short circuit turn has happened inside. This will cause a hot spot and make the varnish very hot and give a pungent smell after a short while. The excess current is often not enough to blow the mains fuse.

3. **Listen** on switch on for cracking, if a valve or ceramic capacitor is cracking over you will hear it, you will also see an upward movement on the plate meter, do not use any more and investigate further.

Common Faults

Mains Fuse blowing in the plug or the back panel of the amplifier. This is the most common fault when an amplifier becomes poorly and needs TLC! When this occurs, the HT supply is breaking down to earth or there is a short on one of the HT components. There is a string of components that could make this happen and you must tackle the string in the correct order as below.

1 Valve checking

Repair the mains fuse with a light fuse, around 2 amps, remove both valves and try switching on. This test will show if a valve has an anode to grid short. If the amplifier switches on then you know you have a faulty 3-500Z.

Next, you can test one valve only to determine which valve is causing the problem. Try one valve in making sure that the free APC (top cap) is well clear of any other components before switching on.

2 Coupling or decoupling Capacitor check

Disconnect the HT lead from the anode RF choke, either at the choke or at the power supply, the choke is the long vertical one between the two valves. This will test for a faulty decoupling capacitor or the main anode coupling capacitor often a (doorknob type) which may have gone short circuit.

3 Power supply Test.

Remove one of the two wires that comes from the HT winding of the transformer to the power supply board, this will disconnect the power supply board, but will test that the mains transformer is ok. If you can now switch on, it means that a component or components have failed on the power supply board. Tests can be done to see if the diodes are ok, also check that the large 200µF smoothing capacitors have not become short circuit.

4 Transformer Test

This now only leaves the mains transformer, the most expensive and difficult component to change, but this also includes mains filtering and decoupling capacitors any of which may have become short circuit. The mains transformer can have several different faults, one is that there is a short circuited turn, sometimes this will not blow the mains fuse, but the transformer will smell and become very hot. Secondly a short circuit between two windings usually blows the mains fuse.

5 Internal Wiring.

Mains wiring in the loom can become faulty but is very rare, the on/off switch can become faulty but rarely blows the main fuse but stops the mains getting to the transformer.

The Valve glows Blue

A 3-500Z will only glow blue when the air seal has been broken, this is often due to excessive heat, vibration or has been dropped. No repair is possible, you need to change it for a new tube! The 3-500Z is like a light bulb, it does not leak air unless it has been damaged.

Unlike ceramic tubes, a 3-500Z does not need 'burning in' to create a better vacuum.

Valve glows Red

It is normal for an Eimac tube to glow red under normal conditions when running near maximum plate dissipation. It will go hot and cold with normal CW and SSB operation. If the valve glows red continuously this is caused by a cathode to grid short, which is one of the most common faults with the 3-500Z. This occurs when the valve loses its grid bias. This means that the standing current is twice the normal and you cannot drive the valve as all the drive is going to earth through the grid. It also shows a high input SWR on your radio transmitter. In time, when the grid starts to age or has been over driven, it distorts and the isolation between the grid/cathode breaks down. Change the valve!

Valve not lighting up

When the valve is running normally you can see the white light from the filament glowing like a normal light bulb. This should be quite bright when everything is normal. If not try the following: -

1 Take the valve out.

Measure with an ohm meter the resistance between the two heaters, pins 1 and 5. This should be very low and under 1 ohm. The fault can be an open circuit filament or a dry solder joint within the 3-500 heater pin. It can be worth trying to re-solder the pin, this can be done by using a large 200 watt soldering iron. Heat the pin, suck out the old solder with a solder sucker, then replace with new high temp solder or lead free solder which will melt at a higher temperature. Lots of 3-500Z have been thrown away when they could have been repaired.

2 Check the heater voltage

Place the amplifier so you can measure the voltage on the valve base. Measure the voltage on the valve base, pins 1 and 5. Measure the voltage on the centre of the valve pins. It should be very close to 5 volts AC. It is common for a bad connection between the valve pin and the valve socket. Take the valve out and clean the valve pins with a very light abrasive, fine sand paper or kitchen pan cleaner. Also remember to clean the valve pin holders on the base, this is best done with a very fine file approximately 5mm across and clean the inside of the pin holder until it is shiny. There is a high current passing through each valve pin. Each valve draws 15 amps heater current so any slight resistance will cause extra heating to the valve pin.

3 Check the heater transformer

Some amplifiers have a common HT and heater transformer but others have a separate heater transformer. If the transformer has failed you can smell it. I have changed several heater transformers in the TL 922. Access to this is by taking the left side of the case off and then it is quite straight forward to access the transformer. These are still available from Kenwood. Other makes of amplifiers use a common HT and heater transformer and are available from the relevant companies in the USA. It is unusual for a heater winding to fail. If a smell comes out of the transformer it is more likely that an internal short circuit between two windings has occurred. If you have 5 volts AC leaving the transformer but not at the valve base, check the connection at each end of the cathode choke as they can look quite normal but over time become an open circuit.

Valve internal Arcing

This is due to the anode and grid getting too close. This is sometimes due to having a bad sideways shock or overheating in the valve anode. When the valve has been overdriven the anode can get very hot and become distorted, it can damage the power supply diodes, inrush surge resistor (if fitted), meter shunts or even the meters. A 3-500ZG is much more robust having a graphite anode rather than the pressed metal type. The support for the anode is only the single pin which comes from the top of the 3-500Z envelope, there are no other supports, so the anode can easily be distorted although the graphite version is much more robust.

Change the valve.

To determine which valve is faulty, try one valve at a time. Switch on the amplifier and set to the 40m band with a CW drive of 50 watts, do all the testing on 40m, the tune and load positions are at their mid range so you will have plenty of compensation in the circuit to still run on one valve. The tune and load positions will change a little from normal. The input SWR will be slightly higher for testing purposes, it will still give you a good idea of the state of the valve, and you should see at least 400 - 500 watts of RF.

What to do after the Valve has arced and the power is still low.

The first item to check is the small choke that goes between the grid of the valve pins 2, 3, 4 and to earth. If this goes open circuit when a large flashover has occurred this will switch off one valve so you only get half the normal power of the amplifier even with a new valve fitted.

High input SWR

The number one reason is when a valve goes grid to cathode short, as described earlier in the article. This will also give a permanent standing plate current with the plate meter showing around 200 mA. This problem switches both valves on because the bias has been lost, both valves are connected in parallel hence both valves are running with no bias.

High input SWR on one band only and low output

1 Input board capacitor

If the high input SWR is only on one band the cause is probably due to a capacitor on the input board breaking down and going short circuit to earth. This can easily be measured with an ohm meter as the whole circuit normally operates at high impedance with respect to ground. A capacitor can go open circuit inside but is unusual. The wires that connect the input Pi circuit could be open circuit due a bad connection on the PCB or a connection to the input switch.

2 The double pole input switch

If one wafer or wiper has got damaged this will also cause a high input SWR. This is rare as this part of the circuit only ever carries 100 watts. This is unlikely to get damaged as this operates at low impedance so there are no large voltages present. If there is a problem it is usually a mechanical linkage problem. e.g. a coupler loose or even the front panel band selector knob slipping.

Input SWR Intermittent

The problem is normally due to the large 2 or 3 pole TX/RX bypass relay having dirty contacts. These are a large type of relay. Usually it is easy to take the cover off and get to the 2 or 3 pole moving parts. Take the return spring off, remove the poles of the relay, clean with a light abrasive such as Brasso or tooth paste. Re -fit the poles back inside the relay.

RF Output on one band only

This is can be when the coupling shaft between the input switch and the Pi or Pi L circuit has become disconnected or loose.

No output except the lowest bands i.e. 80m or 160m.

This is due to severe burning of the large ceramic switch which shorts the parts of the PA coil. The problem can be that the wiper of the switch has burnt away and not making contact with the positions on the PA coil.

Intermittent loss of receive signal.

This is due to the large TX/RX relay having dirty contacts, usually it is quite easy to remove the cover of the relay and remove the moving flapper and clean with a light abrasive. Try not to remove the silver coating on the brass parts of the relay.

Plate Voltage low.

The mains transformer will never go low in voltage. It will either work normally or not at all. If you get a nasty smell this is due to a short circuited turn where the insulation has broken down. Replace the transformer

Low Voltage from the Doubler Power supply

If either a single diode or capacitor goes open circuit you will drop voltage, often also the transformer will hum as it is working only on half wave rectification rather than full wave.

An electrolytic capacitor on the power supply board can develop a dry joint where it is soldered on to the board. Older amplifier solder joints start to dry and become brittle especially with two-wire type capacitors. Larger ones which have fixing screws for their connections do not normally suffer.

Nasty smells!

Electrolytic Capacitor

Electrolytic capacitors have aged and one has become short circuit. This is the start of a chain of capacitors going. An amplifier will work with a short circuit capacitor but will put a higher voltage across the ones remaining. The others will now fail all the sooner with the higher voltage placed across them. When this happens you will usually have fireworks and the mains fuse in the amplifier blows. Change the whole chain for new modern ones - they are much smaller and but very reliable.

Soft Start Resistor

If your amplifier is fitted with a soft start circuit and the relay that shorts the current limiting resistor fails, or any part of the timing circuit, then this leaves the resistor in series with the mains transformer. Your power will be very low and probably the large wattage series resistor will get very hot and even burn up.

Transformers

As described previously, if either the HT or heater transformer develops a short circuited turn then it emits a nasty smell

APC (anti parasitic choke)

Under normal operation of the amplifier the APC is just passive and lets all the RF pass from the valve into the PA tank circuit. If a frequency tries to leave the amplifier that is above the normal operating frequency, the amplifier APC comes into its own and acts as a choke. The resistors in the APC will dissipate the power and will give off a pungent smell. The APC helps stop 3rd and 5th harmonics and improves the stability of the circuit. The combination of the L of the APC coil and the C of the valve create the stop frequency which is normally just above 30MHz. If the coil has been squeezed it lowers the stop frequency. If you operate between 29-30 MHz you find the resistors get hot (and smell). Stretching the APC coil will increase the frequency and put the stop frequency just above 30 MHz. The resistors are normally a combination of one or three giving a value of around 50 ohms (often 3 x 150 ohms 2 watts in parallel)

Maybe NOT the Amplifier!

I have seen many amplifiers that do not have a fault at all. The coaxial lead between the transmitter and the amplifier has an open circuit braid within the PL259 plug. You can still drive the amplifier as the earth return comes through all the other earth points in the shack but not direct through the coaxial cable. Usually the higher SWR is seen worse on the higher 10-20 m bands.

Chapter 8

Fault finding on the 572B amplifier

Fault finding tips and servicing of 572b amplifiers

DANGER WARNING

To start fault finding always be fully aware that inside there is 2400 volts. You must make sure everything is switched off and that the anodes and smoothing capacitors have been fully discharged to earthed. Only do this if you are confident and familiar with high voltage circuits.

This Includes the Yaesu FL2100, B, Z and FL2000B also the Heathkit SB 200 and KW 1000

There have been thousands of 572b amplifiers made from the early 1970s, through to the mid 1990s. A pair of 572b will deliver a nice 600 to 700 watts output or 1200 PEP as some manufacturers would rather quote. I still prefer RMS watts.

They were developed by Cetron as an economical medium power triode. Previous amplifiers in this power class used television line time base valves like the 6KD6 and 6JB6.

Nowadays the only people to make the 572b are the Chinese and badged under different names such as Taylor and Golden Dragon, but all are made in China.

Do from time to time blow out the amplifier. Then clean the high voltage circuits with a tooth brush.

Cooling fans also become dry and can become slow or seized. Clean first and add a few drops of light oil or WD40 into the shaft of the rotor of the fans. The amplifier will still work with no fan running but will run much hotter. It is worth checking from time to time. You will not know the fans in 572b amplifiers have stopped, they tend to be small and very quiet.

Fault Finding

1 Look inside the amplifier and see if both filaments are lit.

2 Look at the plate voltage meter and check the 2400 volts is ok.

This can tell you:-

a) No voltage could mean the transformer or voltage doubler is faulty or just the fuse on the back panel has blown.

b) If the voltage is around 1500 volts, it could indicate that one of the electrolytic capacitors has gone open circuit and the others will have been stressed.

Check the PSU PCB for burns and open circuit tracks.

c) If the voltage doubler board is lower than 1000 volts, suspect one of the diodes in the chain.

3 Listen for arcing on switch on, this could be from a valve or an electrolytic breaking down.

If from the valve! Change it before any other damage happens. You can often see the arcing from inside the valve.

If you think the arcing is from a smoothing capacitor, it is always best to change the whole chain of capacitors as you normally get a chain reaction. The capacitors are either 450 or 500 V DC working. When one breaks down it increases the voltage on the others in the circuit. Then one row of the capacitors will completely go short circuit and eventually blow the mains fuse in the back of the amplifier. Change all for more modern ones. They will be smaller and more reliable.

4 Switch the amplifier on, key the amplifier and observe the plate current.

If the standing current is higher than normal and you have a higher than normal input SWR on your radio, suspect one of the valves. The most common fault that occurs is a grid to cathode short. Check between the two larger pins 1 and 4, which is the heater, and pin 3, the grid. This can sometimes be permanent but also can happen when the filament comes up to temperature. Change one or both valves and try again.

5 Low output around 300 watts.

This is often when a valve has flashed over internally. The flash-over is from anode to grid and this in turn takes out the grid resistor which is 33 ohms. This can be found in the small box on the back of the amplifier (FL2100s). The 33 ohm resistor is the first thing to go open circuit when a valve has flashed over. It takes the full HT voltage to earth through the 33 ohm resistor. It is easy to check with an ohm meter. Low output power can also be when the valves have aged and have become tired. When the glass goes dark it shows that the valves are ageing. Like a fluorescent tube, the emission reduces and maximum power can be as low as 400 watts or less. Change both valves.

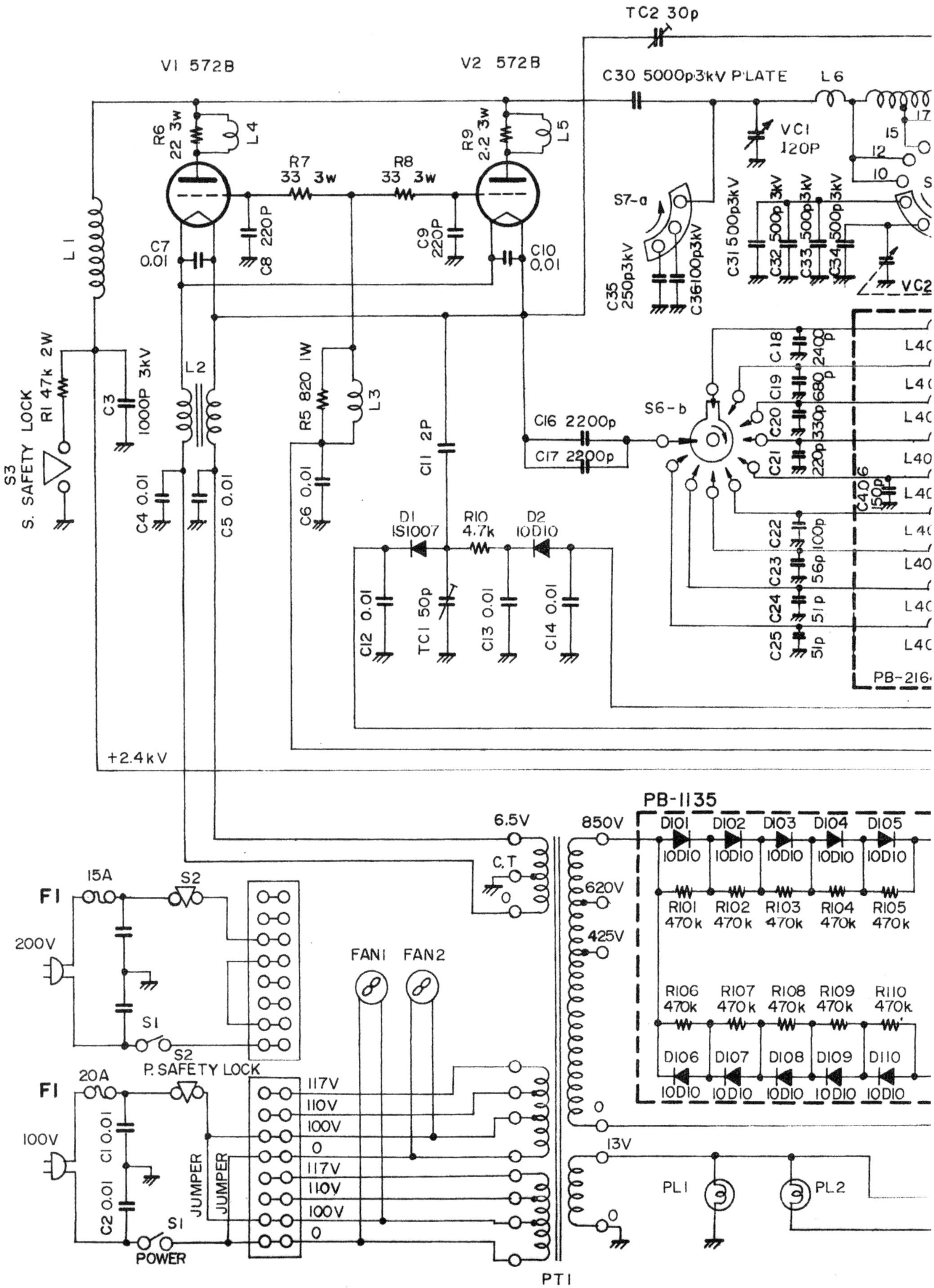

TC2 30p
V1 572B
V2 572B
C30 5000p 3kV PLATE
L6
R6 22 3w
L4
R9 2.2 3w
L5
R7 33 3w
R8 33 3w
VC1 120P
17
15
12
10
S7-a
L1
C7 0.01
C8 220P
C9 220P
C10 0.01
C35 250p3kV
C36 100p3kV
C31 500p3kV
C32 500p 3kV
C33 500p 3kV
C34 500p 3kV
VC2
R1 47k 2W
C3 1000P 3kV
L2
R5 820 1W
L3
C11 2P
C16 2200p
C17 2200p
S6-b
C18 2400p
C19 680p
C20 330p
C21 220p
C406 150p
C22 100p
C23 56p
C24 51p
C25 51p
L40
PB-2164
S3
S. SAFETY LOCK
C4 0.01
C5 0.01
C6 0.01
D1 1S1007
R10 4.7k
D2 10D10
C12 0.01
TC1 50p
C13 0.01
C14 0.01
+2.4kV
PB-1135
6.5V
C.T
0
850V
620V
425V
0
13V
D101 D102 D103 D104 D105
10D10
R101 R102 R103 R104 R105
470k
R106 R107 R108 R109 R110
470k
D106 D107 D108 D109 D110
10D10
F1
15A
S2
200V
S1
FAN1
FAN2
F1
20A
S2
P. SAFETY LOCK
100V
C1 0.01
C2 0.01
JUMPER
JUMPER
117V
110V
100V
0
117V
110V
100V
0
S1
POWER
PL1
PL2
PT1

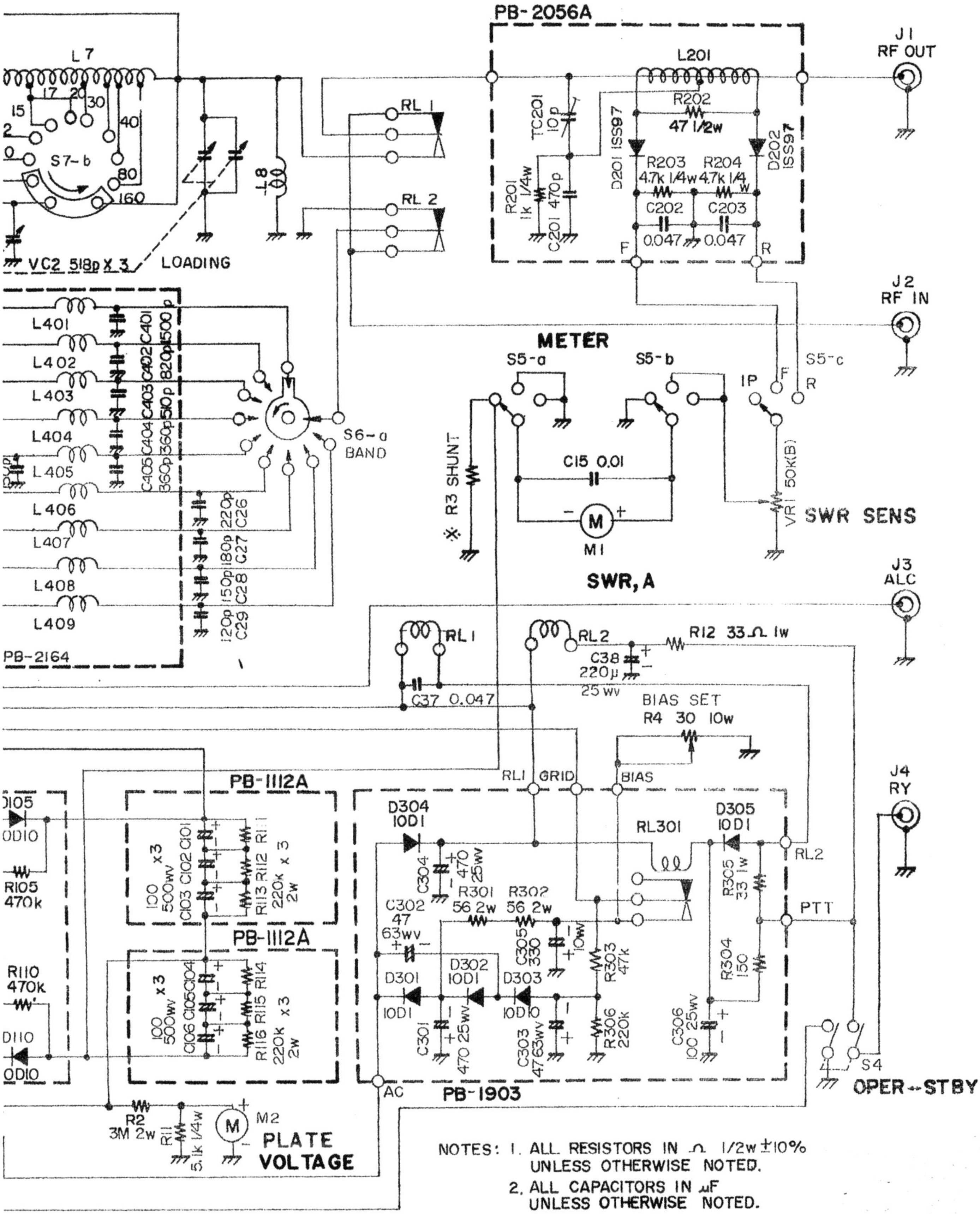

NOTES:
1. ALL RESISTORS IN Ω 1/2w ±10% UNLESS OTHERWISE NOTED.
2. ALL CAPACITORS IN μF UNLESS OTHERWISE NOTED.
3. ※ VALUE IS NOMINAL.

6 High through SWR even when the amplifier is switched off

This is due to dirty contacts on your TX/RX relay. Found on the underside of the amplifier, follow the wires from the input and output SO239. If possible take the relay pole out and clean the contacts. With most relays you can remove the cover, take off the small spring that returns the poles back to the rest position. Use a light abrasive or very fine sand paper to clean the contacts. Before replacing make sure they are all in line.

Caution!!! Before pulling the amplifier to bits, check your PL 259 leads are OK.

7 High input SWR on one band.

The cause could be that the input circuit board has a faulty capacitor. Check that particular band capacitor on each side of the coil is above earth.

Check continuity of the coil from the input wafer to the output wafer. There should be continuity.

8 No keying of the amplifier!

This fault is rarely the amplifier, it is more often that the PTT lead to your transceiver has been pulled and broken.

Also!
If you have changed your radio or you have done a software reset. The TX relay will have defaulted back to being switched off. Reset the position in the menu of the radio.

9 The amplifier only works on one band.

The amplifier has a coupling shaft between the PA ceramic switch and the input coils at the back of the amplifier. The coupler can become loose and leave the PA switch fixed on one band. Realign and tighten the coupling shaft.

10 The fuse blows on switch on.

This can be a valve having an internal short between grid and anode. Remove both valves and replace the mains fuse. Check if the fuse now holds. Make sure the APC on top of the valves are away and not touching any other components before you try switching on.

If the fault remains it is due to the doorknob capacitor on the top of the anode choke going short circuit, it is 500pF at 3 kV. You can often see a crack in the capacitor or a black mark on one side of it.

If this does not fix the fault it must be a decoupling capacitor on the bottom of the anode choke L1, the capacitor is 1000pF at 3 kV.

Chapter 9

Amplifiers made by Linear Amp UK
Explorer Hunter
Ranger 811
Evolution of the Challenger
Evolution of the Discovery

The Start of Heatherlite and Linear Amp UK

Evolution of the Explorer and Hunter

The Heatherlite Explorer was named after my first wife Heather. My first production amplifier was the Explorer back early 1980s after building the classic 4-400 amplifier described in the ARRL hand book. At the time 4-400 valves were readily available at most radio rallies as surplus, often brand new for as little as £20 each. The matching bases were also available.

My friend John G4FCT and I had the handout leaflets for the Kenwood TL922 which at the time was out my pocket. We both admired it and we decided to try to copy it as near as possible. John later purchased a new TL922. We bought square-type box section tubing with 3 way corners to make the cabinet. We got all the side panels made and off we went. The amplifiers were eventually finished and produced around 800 watts on the amateur bands. The valves were really under run and even on RTTY the valves did not get hot at full output power. The 4-400 anodes were made of graphite before the 3-500ZG graphite version arrived. The 4-400 were used for military transmitters and ship radio transmitters, with their high gain they could be used with a low level of drive.

I remember putting my first advert for the Explorer in RadCom. At the time I was still a young Radio Amateur with a lot to learn. There was a net on 80m on 3.680 with a group of Headteachers and people who had their own businesses. This net was at a level that I could only aspire to! They had built most of their own equipment including G2DAF receiver and transmitter. The person who was the net leader G3AOO rang me and said he would like to buy one of my new amplifiers! He bought it and I heard it being used every night. I was quite proud that I had gone up in the world with his very nice comments about this British built Explorer Amplifier.

I made around 500 Explorers and Hunters. One Explorer was in use for many years at GB2SM the Science Museum in South Kensington. Eventually the station was closed down as new methods of communications went on display - perhaps amateur radio was of no great interest to the public any more! I also made a marine version some of which went to the Falklands and others went to the Air Sea rescue HQ in Edinburgh working on 4 and 8 MHz.

My First Explorer

This Explorer below was one of my first production units with a brown front panel and beige covers. I wanted to be individual, maybe like Heathkit and their green cabinets.

I first started going to radio rallies selling mobile microphones, this was before it was illegal to drive holding a microphone while driving. I met Hedley Morgan Smith, High Wycombe, Buckinghamshire, selling very nice cabinets. I told him of my wish to start making amplifiers so I did drawings (of course before Cad became available), bearing in mind I had never done professional drawing before! I chose the colours as he could also do powder-coating paint. For a period of about eight years Hedley made all my cabinets for the Explorer as well as the Hunter and 2m Explorer.

Later a friend from Lancashire G4YSN who is now GM4YSN who then was much closer in distance approached me to see if they could give me a quote for cabinets. I agreed and now all the cabinets are still made near Chorley Lancashire by Bonney Greenhalgh and Company, who have been so helpful over all my time making amplifiers.

The top inside of an early Explorer 1000

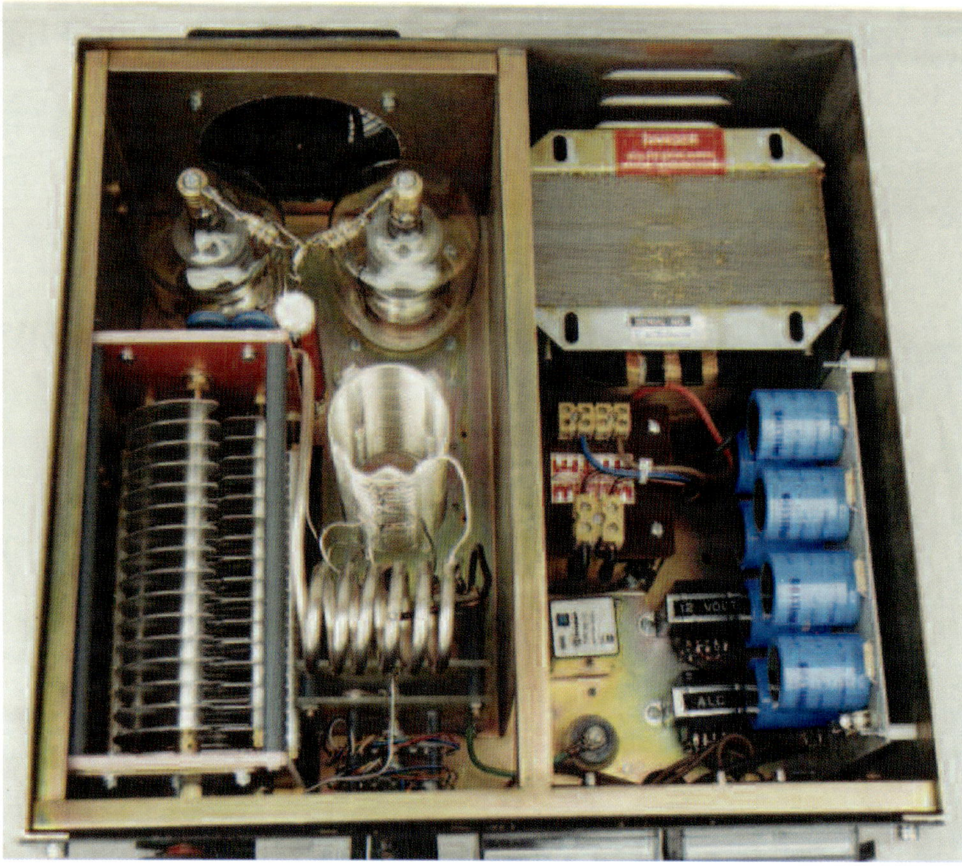

The picture above shows one of the first Explorers in its anodised steel case. You will see the two large iron core transformers for the HT and heater supplies. The 2 x 3-500Z are in the top left corner, I even turned and made the brass top connections for the top of the 3-500Z.

The large tune and load capacitors were made by Tony Johnson of Cap Co Electronics Ltd. Later Lloyd, Tony's partner, continued to make the capacitors for me until I finished manufacturing amplifiers in 2008. The large ceramic switch which you cannot see was made for me by Radio Switch Corporation in New York.

As much as possible at Linear Amp we made all of our own components including the silver plated tank coils for the PA, which we even silver plated ourselves.

The tuned input boards were all hand made. I used a lathe to wind the anode RF chokes and a strong hand to make the cathode chokes on a former.

To begin with I used Eddystone formers for the 40 to 160m coils. When they ran out I went to a local pottery in Hornsea who made the ceramic formers for me. Later I went on to use Airduct coils from Annecke in Germany and B and W coil stock from the USA.

In Louth, Lincolnshire, I went to see Douglas Transformers, a well known transformer manufacturer. They made me the first prototype and later ordered in them 10s for both the HT and heater transformers, quite an investment for me at the time to get the price down along with the cabinets and valves etc.

Quite a few people have copied my amplifiers, that is fine! But it takes the biscuit when someone copies it and puts the article in RadCom then gets paid for the article. One who copied the Explorer and another who copied the Hunter 750, in every dimension and circuit.

But I guess Imitation is the greatest form of flattery!

The Explorer and Hunter ceased production after Eimac stopped making glass tubes. I continued for a while with the Amperex 3-500Z but the reliability was poor and they admitted to a problem in the production. This was before the Chinese ones became readily available.

I do not have a picture of my very first Explorer which was the prototype. It was bought by Peter G4HHH in Whitby and was used for many years. He bought another when I changed the design for the final time.

The Explorer 1200

The last Explorer 1200 with WARC bands and the charcoal look

Inside of the Explorer 1200

Top right is the 1.2 kVA toroidal transformer

Top left is 2 x 3-500ZG.

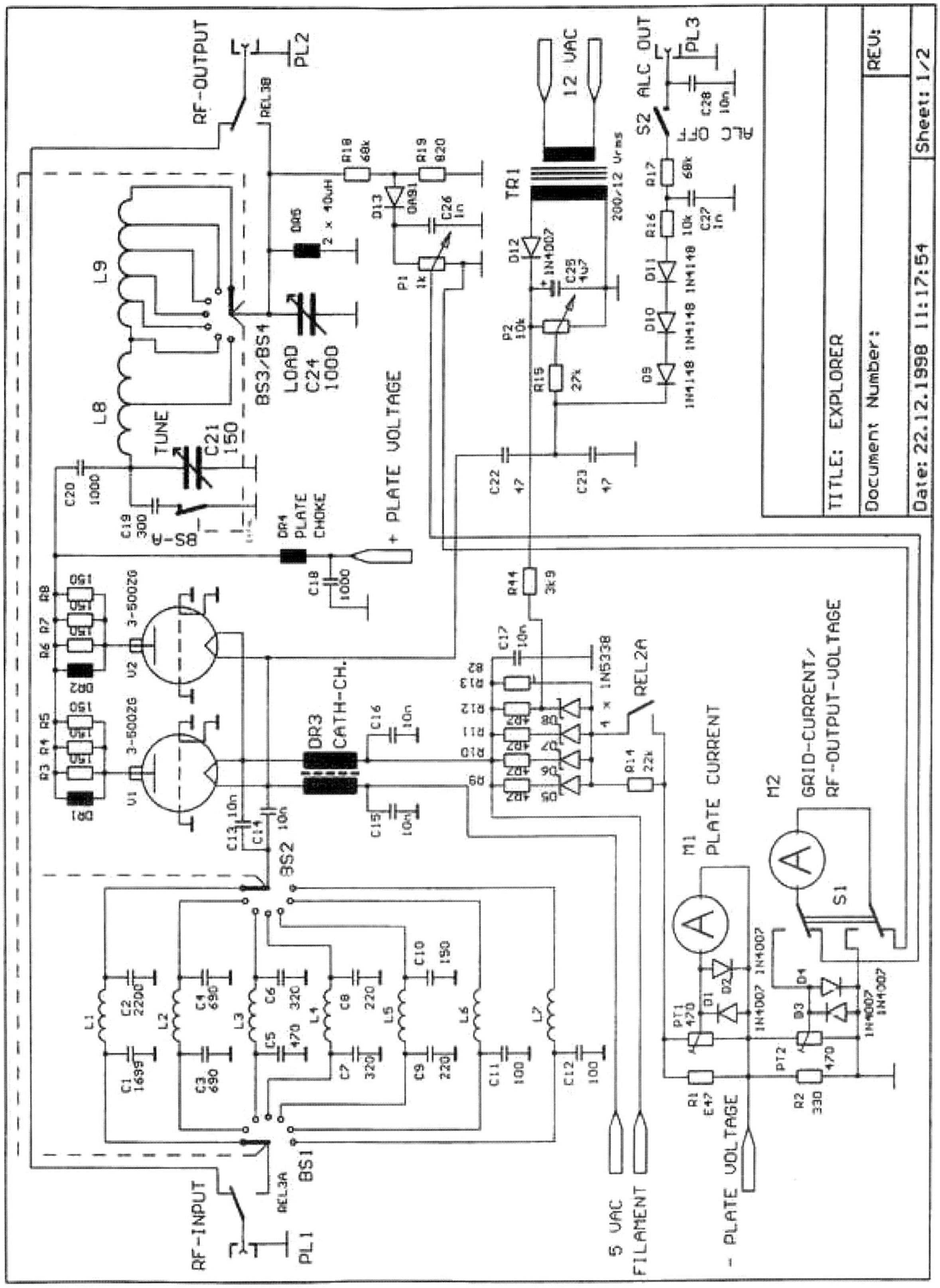

RF section of the Explorer

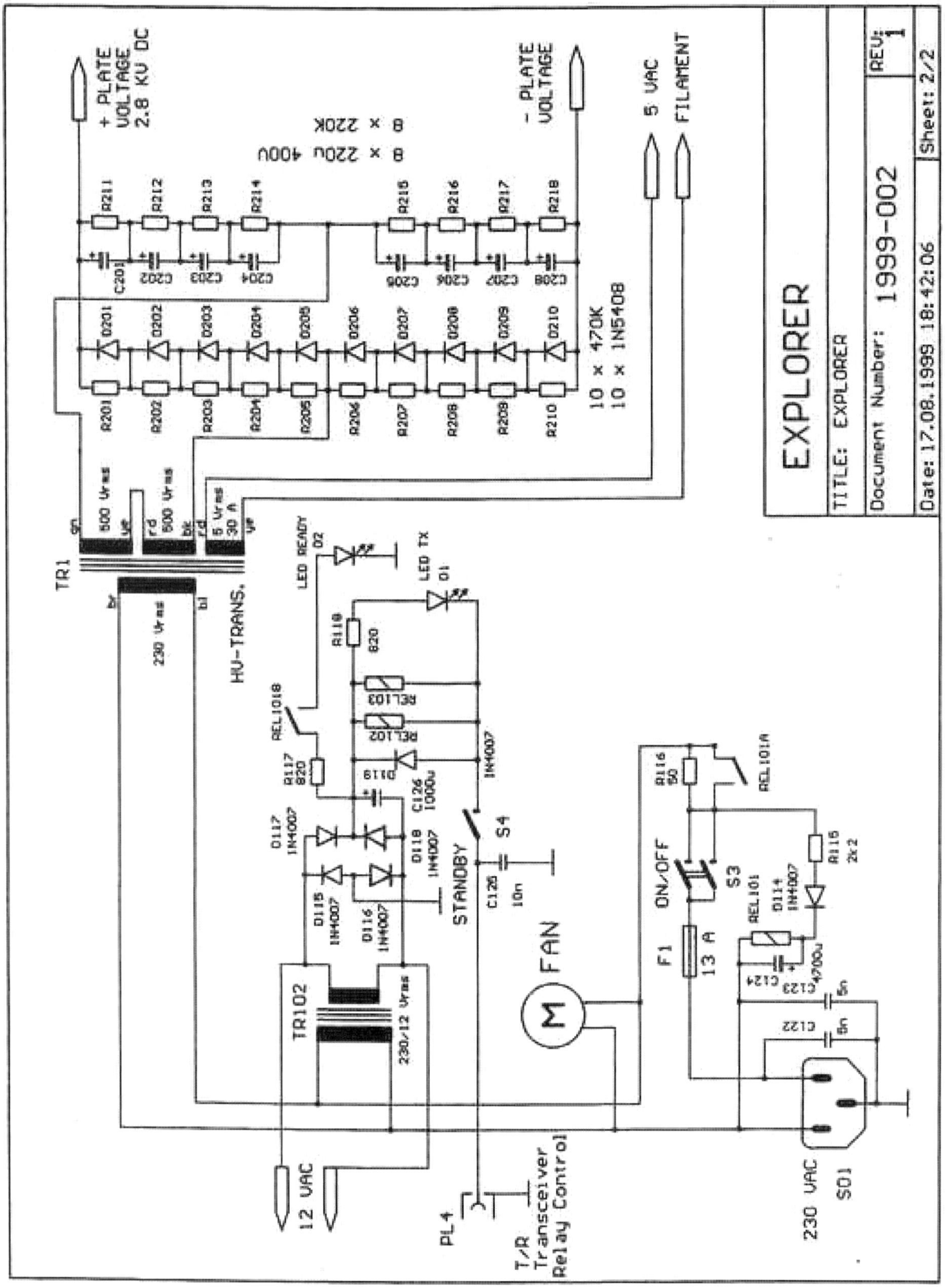

Explorer power supply unit

Linear Amp Hunter 750

The Hunter was made after the Explorer had been on the market for some time. It was a cheaper alternative to the Explorer having only a single 3–500Z but it would still produce the UK legal limit of 400 watts. The Hunter originally only covered 10m to 80m but later ones had all bands including WARC and soft start was fitted as standard. The Hunter was also made as a single band 6m model called the Hunter Six. This stopped when the GS31 6m Discovery came out.

The Hunter to my knowledge was the first Amateur Radio amplifier to use a toroidal transformer for both the HT and the heater supply.

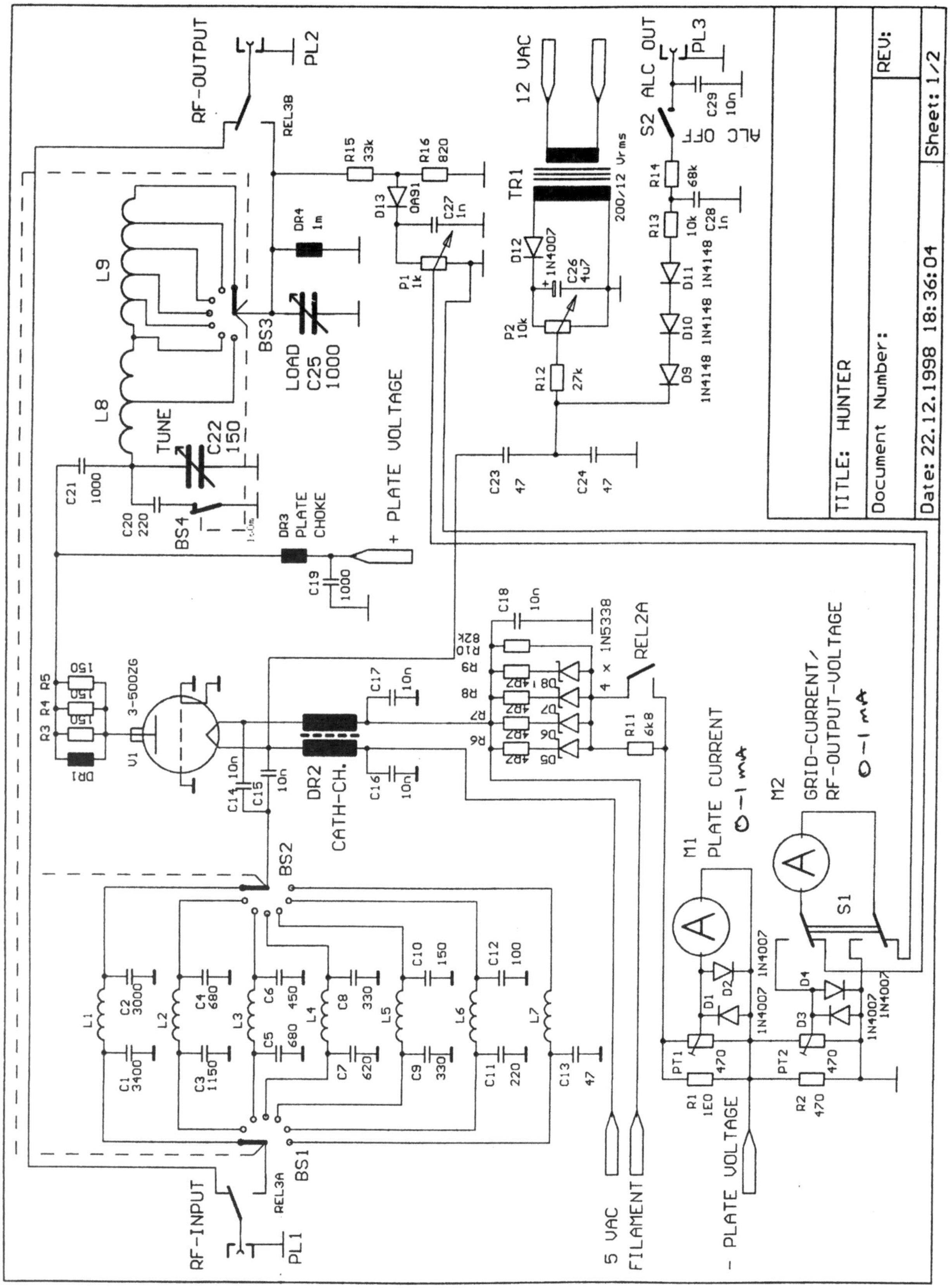

RF section of the Hunter © G3ZRS

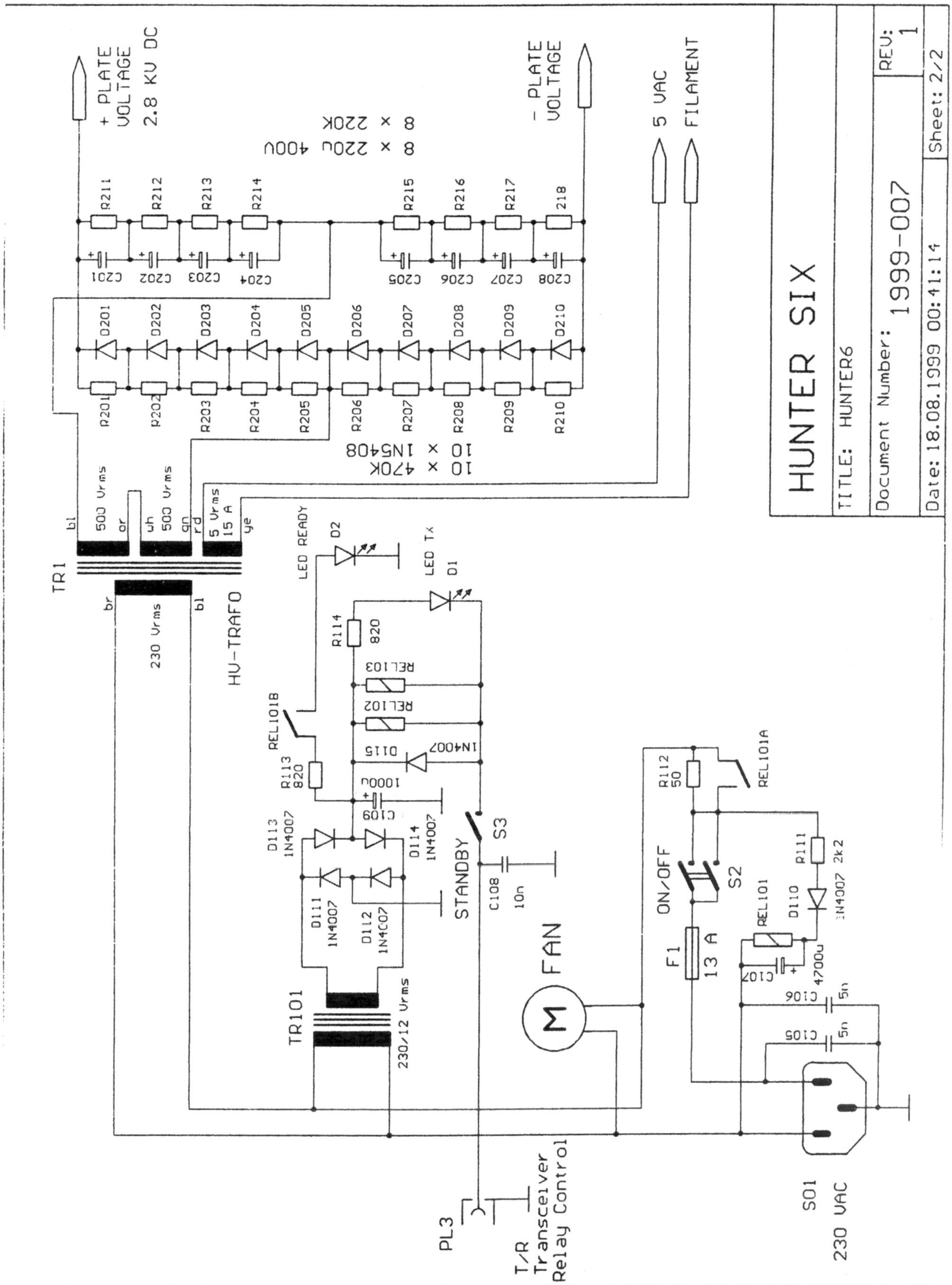

Power supply section of the Hunter © G3ZRS

Linear Amp UK RANGER 811H

The Ranger 811H was to fill a gap for a low priced legal limit amplifier using good quality components. The 4 x 811 valves will deliver 800 watts so the amplifier running at half power would give a comfortable 400 watts. I had been impressed with the Collins 30L1 so I decided to try and make a similar one at an affordable price and add 160m, soft start and the WARC bands. The need for the Ranger 811 came because my Hunter 750 which had been a popular low power version of the Explorer 1200 became difficult to make when Eimac ceased manufacture of the 3-500Z. I bought Amperex ones for a while but the reliability was poor and the Chinese ones were only just starting to become available.

I modified a Hunter case to try the 4 x 811 vertically mounted and the prototype was made! We sold direct to the public and were also available from Waters and Stanton and Martin Lynch in this country and other shops around Europe. It had several improvements over the years mainly with the control board but the outside always remained the same. Over 500 Rangers have been made and it became my most popular amplifier.

The Radio-communications Agency (now OfCom) bought two Ranger 811H for the newly equipped Marconi Demonstration Station at Poldhu, Cornwall, which still operate on a very regular basis as a perfect demonstration of Amateur Radio and the wonders of the early transmissions of Marconi.

The Ranger 811H uses four vertically-mounted 811A valves to produce 800W on the Amateur HF bands. The Pioneer 572b version will give 1200 watts from 4 x 572b with a higher HT of 2400 volts.

Along with all my models, the Ranger/Pioneer has a toroidal transformer providing the power into a voltage doubler board.

To aid tuning even more there are 6:1 reduction drives on the tune and load capacitors. These capacitors are a double module exclusive to Linear Amp UK Ltd.

The valves are cooled by a 120mm, low noise, Papst fan which gives excellent cooling but is so quiet you will hardly know it is running.

Front panel features include two backlit, flush-mounted meters which show plate current and grid current. The grid current meter also doubles as a relative RF power O/P meter, selectable on the front panel. Fully adjustable ALC is also controlled from the front panel.

Later models had an upgraded control board with flash-over protection and a circuit breaker instead of fuse on back panel. The Ranger can take 4 x 572b which are more robust but you will not get any more than the 800 watts that you would get out the 4x 811. This is due to the gain being the same - to get more power the HT would need increasing from 1800 v to 2400v which the 572b will happily take. There are no changes required when changing from 811 to 572b. Both valves have two large and two small pins so just make sure you put the two larger pins in the correct holes of the valve base.

Bottom view of the Ranger 811H

Specifications Ranger 811H

Output power	800W CW or SSB, 400W RTTY
Gain	10 dB nominal
Power requirements	230V AC at 8 Amps, 50/60Hz
Cooling	Papst 120mm axial fan
Frequency coverage	amateur bands 1.8 – 29.7MHz
Valves	4 x 811A, vertically mounted
Harmonics	-50dB typical
Intermodulation	better than -35dB
DC Voltage	1800V DC
Metering	Plate current, switchable Grid current RF O/P
Protection	Primary AC fuses and soft-start on switch-on
ALC	Front panel adjustable ALC control
Dimensions	14in wide x 9.5in high x 16in deep
	355mm x 240mmx 405mm
Weight	25kg

Top View of the Ranger 811

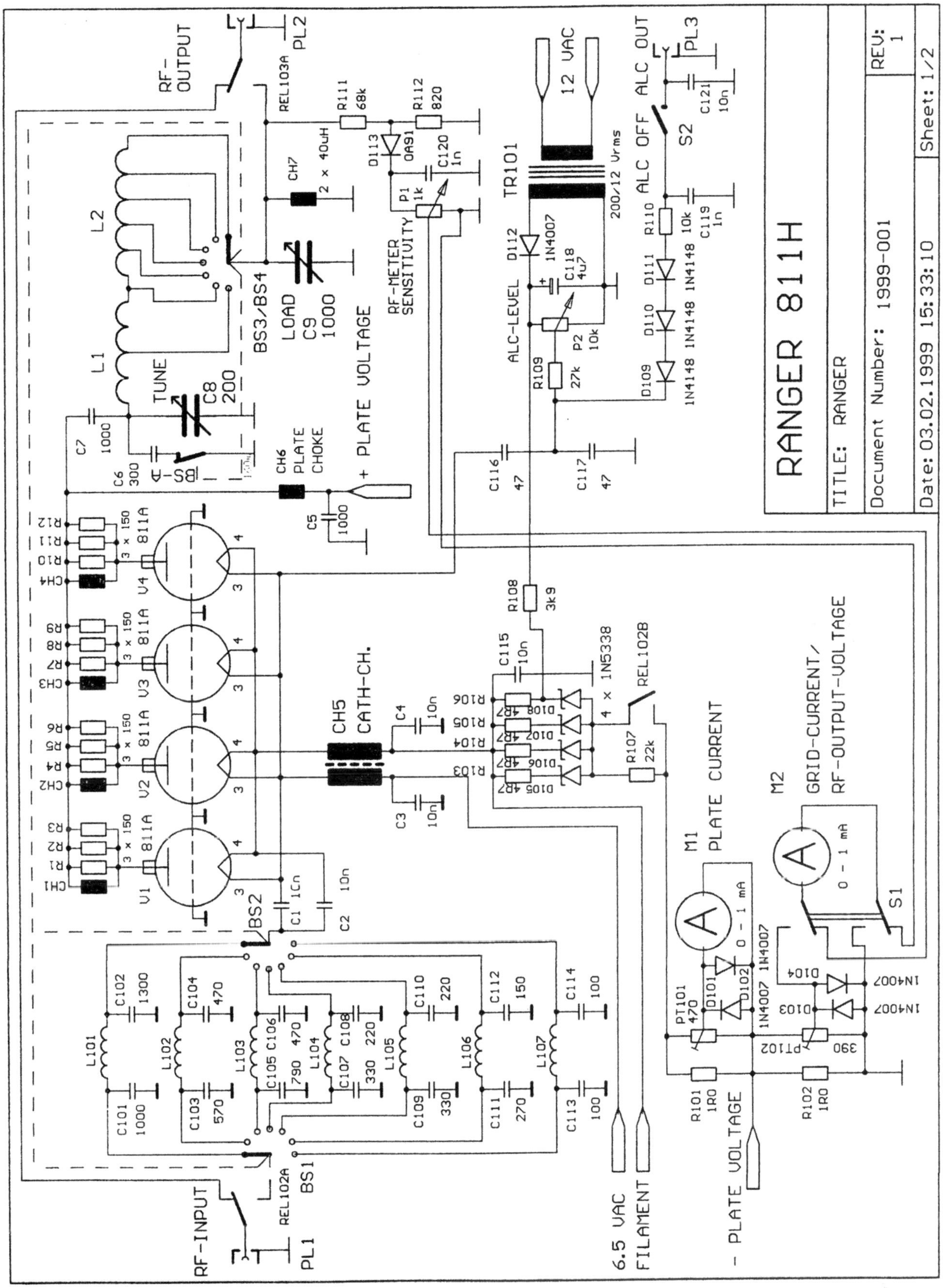

Ranger 811H RF circuit ©G3ZRS

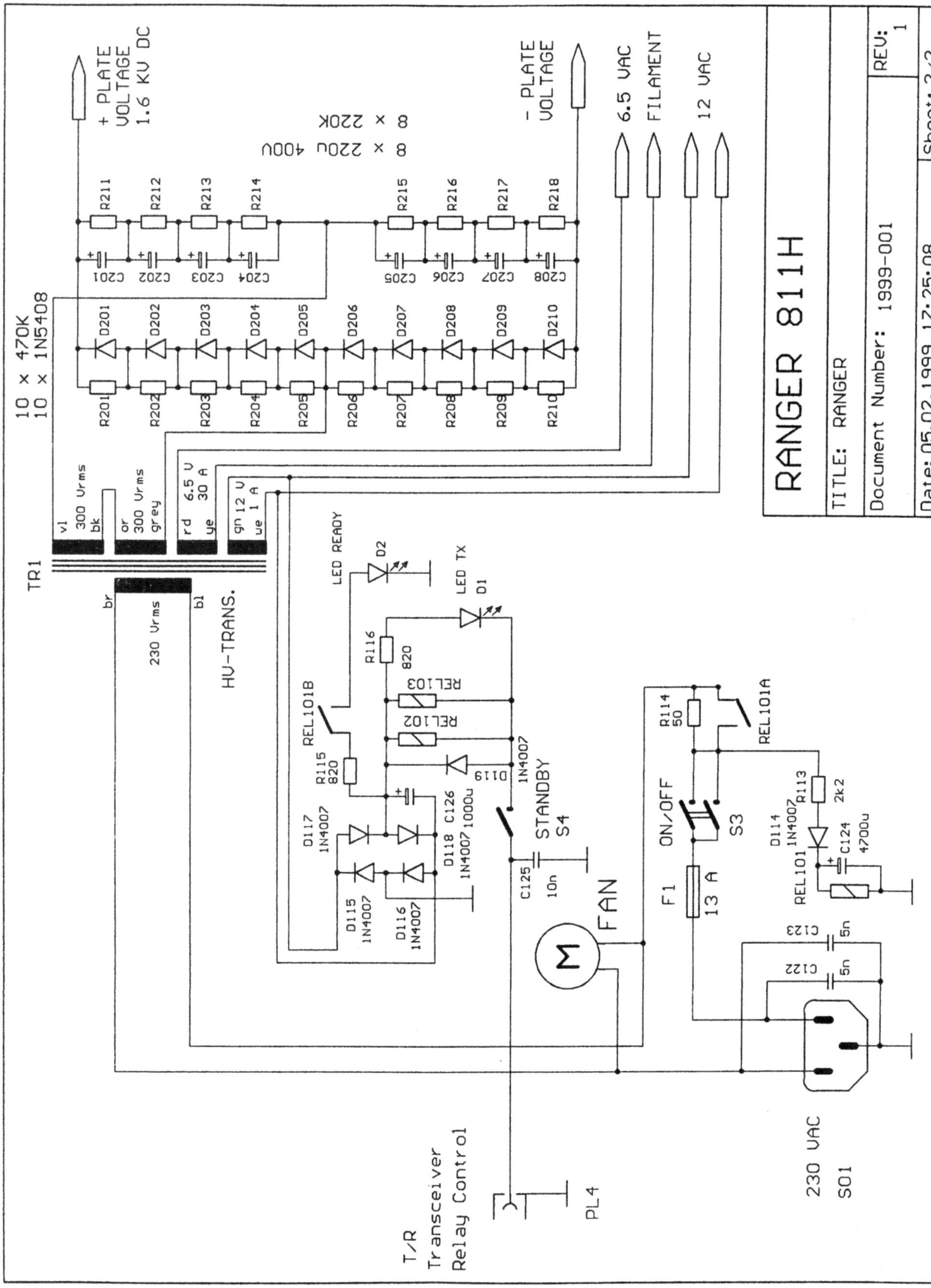

Ranger 811H PSU circuit ©G3ZRS

The Evolution of the Challenger

At the time there was a growing demand for amplifiers that would produce more than just the normal 1kW. The power in the USA had gone up from 1kW to 1500 watts and other European countries had also had got higher power licences. There was demand from the low band Dxers were each dB counts in the grey line time. On 10 to 20m if the band is open you will be heard at some level and your strength will be determined more by your aerial gain.

On the low bands, particularly 80m, working DX can be a big challenge and probably the band where the aerials take up the most land. If you do not have the room for a 4 square, then aerials are very much a compromise and every watt of power helps get that far away DX station. For example, working the Pacific islands takes every dB you can find on 80m and often the signal only ever reaches a few dB above the noise level. American amplifiers were being imported, namely the larger Ameritron AL1500 and the Ten Tec Titan, so I thought I would also add a British alternative! Hence the Challenger I was launched.

Linear Amp UK Challenger I

My first Challenger I would give a very clean 1500 watts. It used two 900 VA transformers which were used originally in the Hunter 750. The two transformers were connected in parallel to give current capacity 2 amps at 2500 volt. Note the external fan which was later changed to a flat pack blower mounted inside. The two transformers were later changed to a single large 2kVA toroidal transformer.

The choice of the 3CX800A7 and later the 3CX1500A7 was due to its tested reliability in the medical field as well as radio communications. Both valves have a fantastic gain compared with any glass triodes, this is due to their special high μ internal beam focusing design.

Challenger II with 2x 3CX800A7   1500 watts 10 to 160m

Both Alpha and Tec-Tec used the 3CX800A7 in their 1500 watt amplifiers. Their approaches were quite different. Ten-Tec made their power supply to go on the floor in a separate box. Alpha made their amplifiers in one box and that is the method I preferred. Having a floating 2500 volts connecting cable can be somewhat dangerous, especially if the dog decided to chew it!

The picture above shows the Challenger II with the fan mounted under the transformer which helps the cooling of the transformer as well as cooling the two 3CX800A7 valves.

Both Challenger I and II amplifiers would deliver a comfortable 1500 watts with less than 80 watts of drive. The progression from Challenger I to II was the new type blower giving a higher pressure of air, PTFE chimneys to funnel the outgoing air from the two valves as well as the changes to the power supply and control boards to give more protection to the ceramic triodes.

Price eventually made it more viable to change to the 3CX1500A7 and gave the same output as 2 x 3CX800A7. The 3CX800A7 uses a costly special 11 pin base made by Johnson. The 3CX1500A7 uses the same base as the old QQV06-40 valve at a reasonable cost.
The 3CX1500A7 is also made in China where the 3CX800A7 never have been in any quantity.

Challenger II circuit using 2 x 3CX800A7

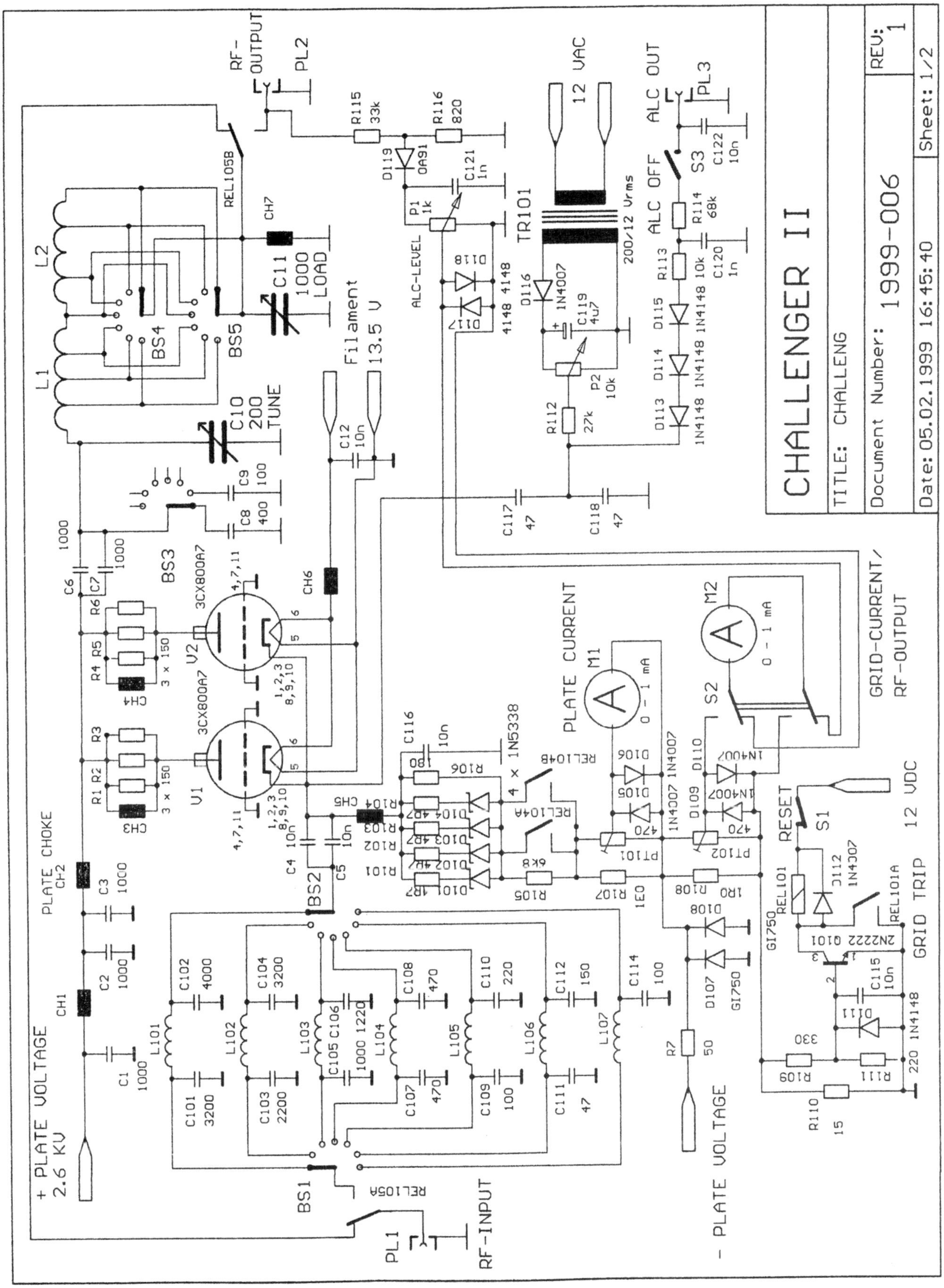

Challenger circuit ©G3ZRS

Power supply of the Challenger II

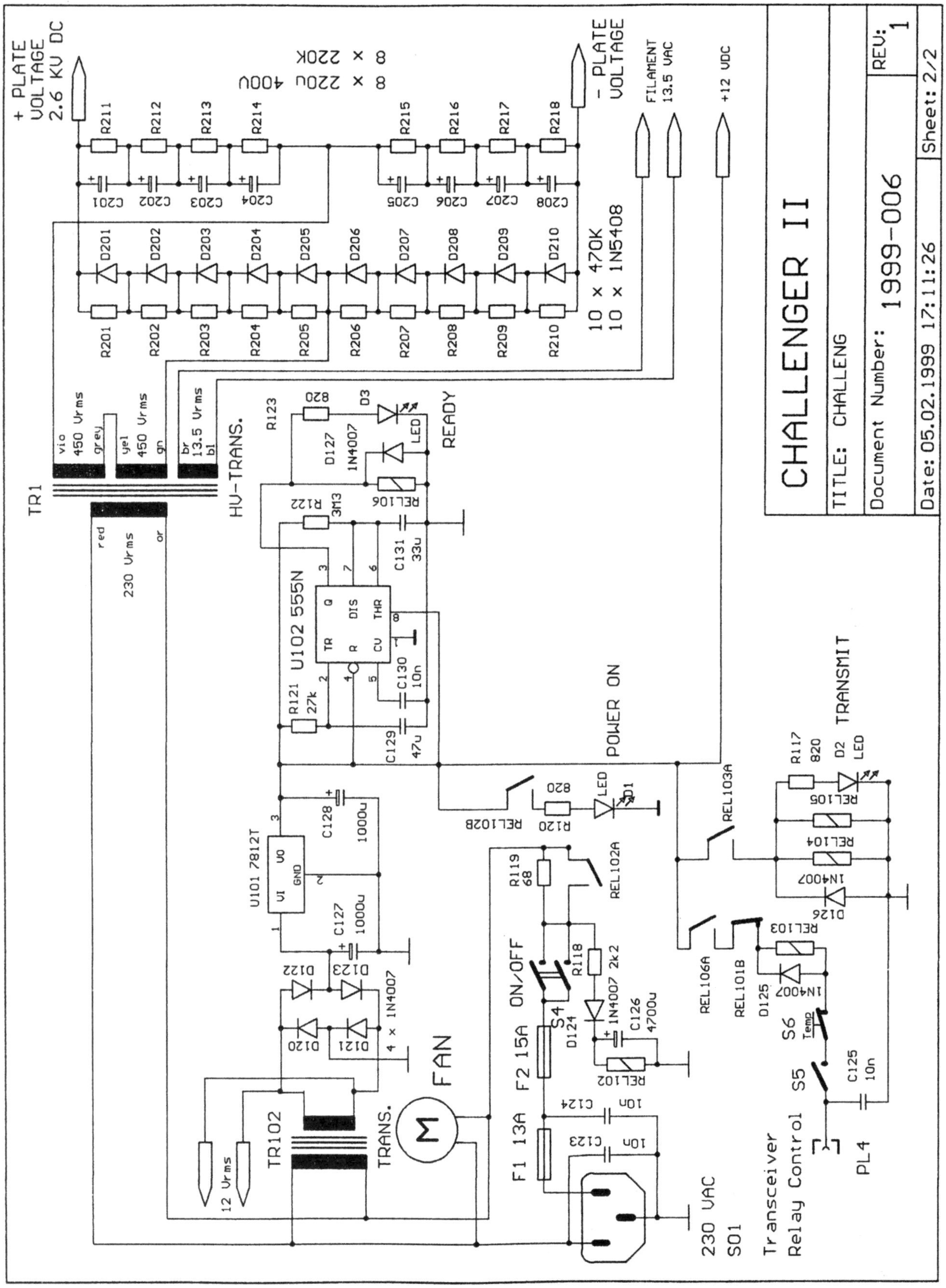

Challenger PSU ©G3ZRS

Linear Amps UK Challenger III

The Challenger III was the first commercial HF amplifier ever to use the Russian GS35b ceramic triode valve. This is a very rugged valve, which is capable of producing high power for long operating periods and the price will not break the bank.

It uses a massive 2.2kVA toroidal transformer producing 3600V after the voltage doubler board. Cooling is forced air from a Papst flat pack blower located just under the transformer.

The Challenger III provides protection for grid over-current, plate over-current, over drive and internal valve flash-over. It also provides soft start, warm up timer, sequenced relays and high speed switching relays.

The Challenger III will deliver 1500 watts on all the HF bands 10-160m with separate band switched positions for the WARC bands. The tune and load capacitors are a large double module unique to Linear Amp UK

Challenger III using the Russian GS35b

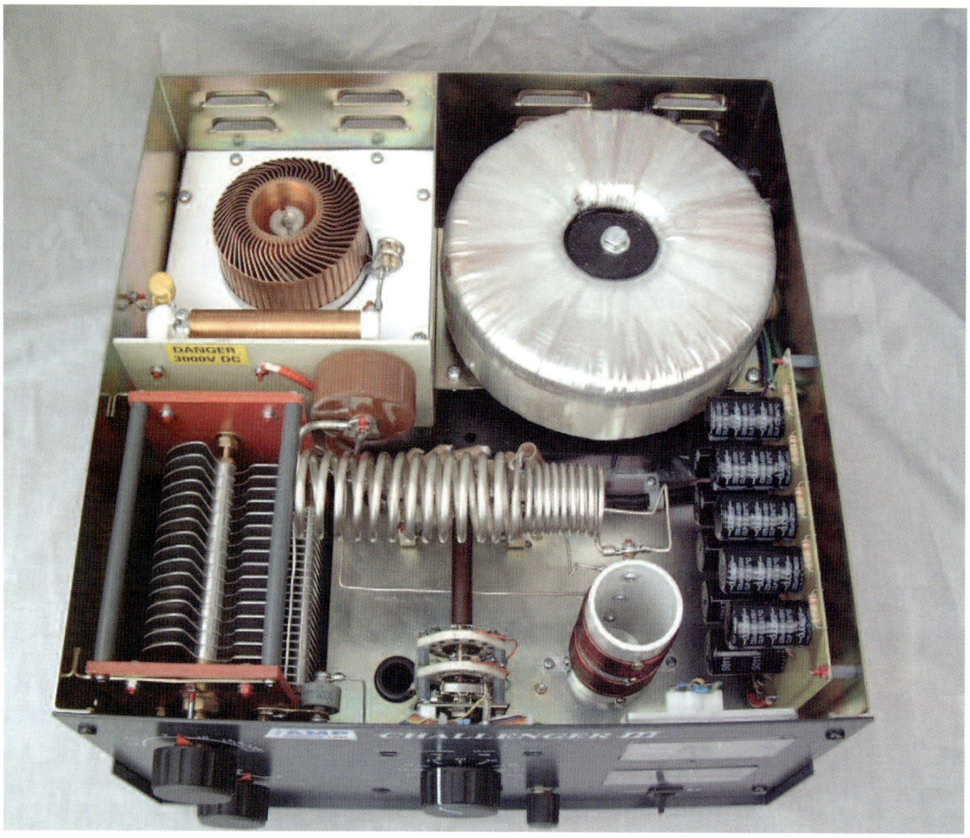

Specifications

Output power	1500W SSB or CW, 1000W RTTY
Gain	13dB nominal
Power	230V AC at 20 Amps, 50/60Hz
Cooling	Forced air cooling
Frequency coverage	Amateur bands 1.8 – 29.7MHz
Valve	Single GS35b ceramic triode
Harmonics	-50dB typical
DC voltage	3600 VDC
Metering	Plate current, switchable Grid current / RF O/P
Protection	Primary circuit breaker, grid current, soft-start start up timer
Dimensions	16in wide x 9.5in high x 17in deep
	410mm x 240mm x 475mm
Weight	29kg

Challenger IV

The Challenger IV using the 3CX1500A7.

The picture below shows the underside of the Challenger IV using the 3CX1500A7 and the G3ZRS/G3TEU control board, the tuned input circuits are mounted down the middle of the amplifier.

Control board as used in the Challenger and Discovery Amplifiers

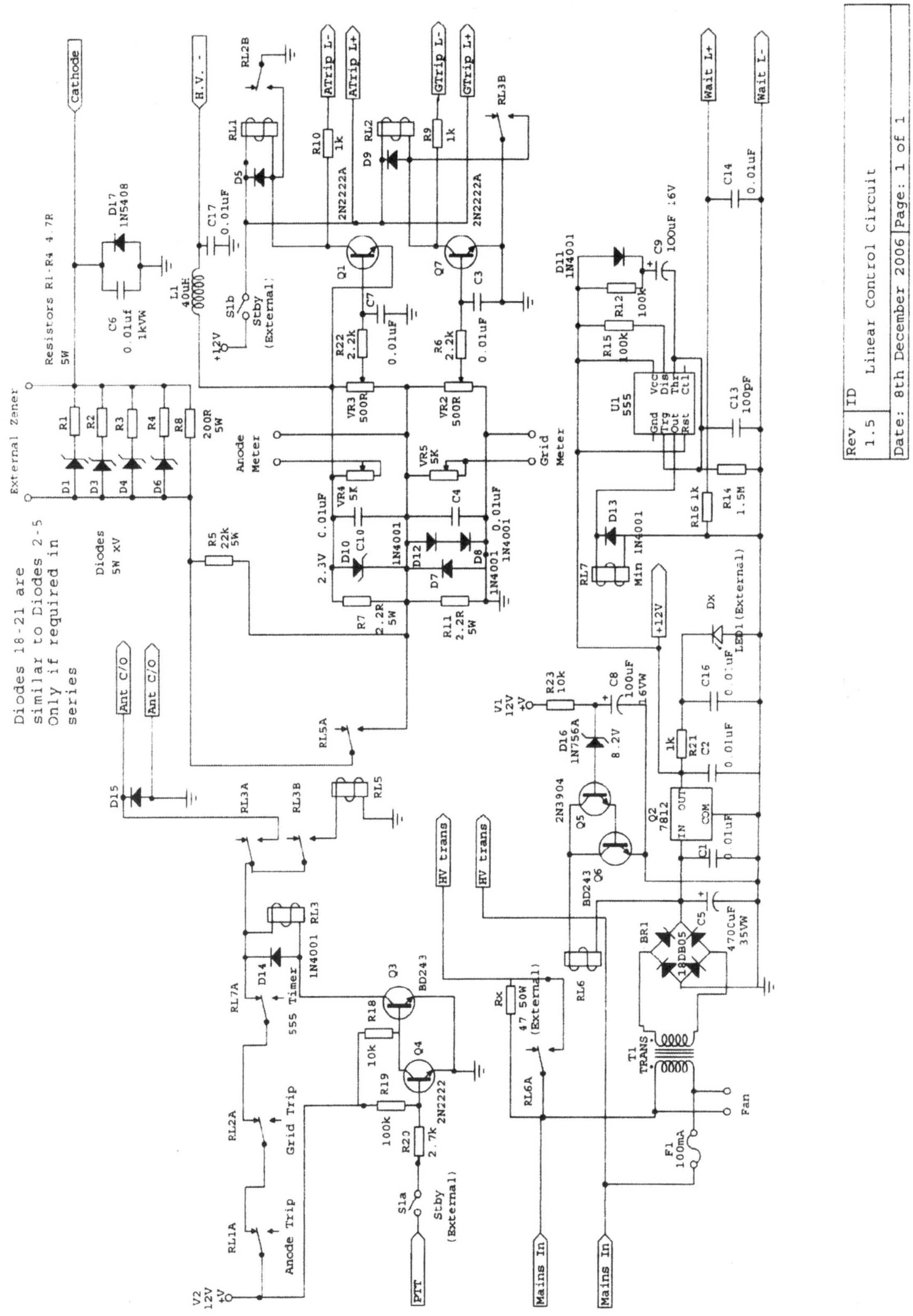

Linear Amp UK Control board ©G3ZRS

This was my last control board which has soft start, low current relay switching, start up timer, grid and anode over current protection and sequenced relays as well as the outputs to meters and LEDs. It is almost completely RF immune.

Over 15 years, 500 Challengers were built and sold mainly around Europe and the Middle East after exhibiting at Friedrichshafen and other Ham Radio exhibitions in France and Holland.

I am glad that Paul Cullen carried on building the Challenger IV down in Fleet, Hampshire with some of his own improvements and later the move to Roger Banks, GW4WND, at the DXshop in Wales.

I think now with the ever increasing need for more frequencies and power, I would be developing a tetrode amplifier such as the Russian GS23b or the Chinese 4CX1000A.

The Evolution of the 70cm, 2m, 4m/6m Explorer and later the DISCOVERY

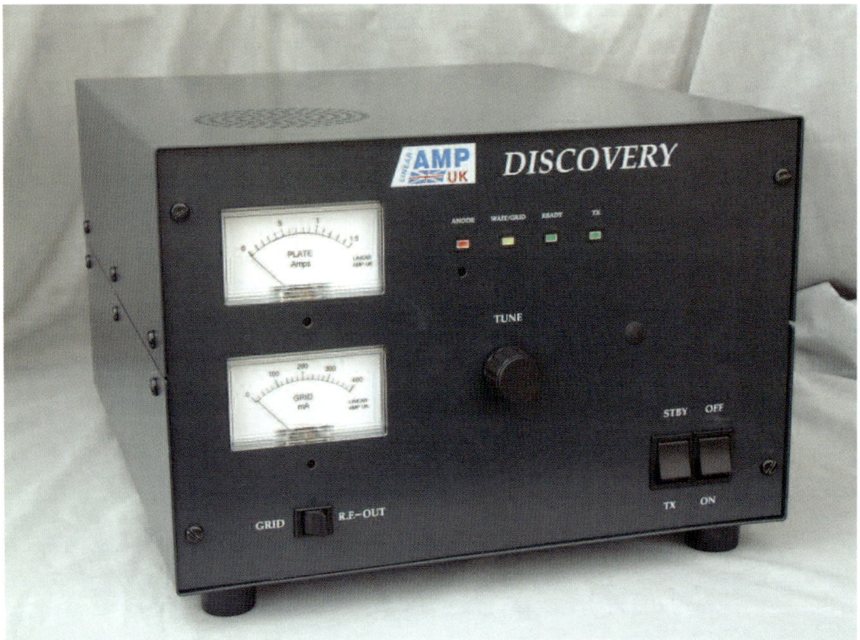

In the 1970s and 80s my first VHF amplifier was the 2m Explorer using a single 4CX250b or 4CX350b giving around 300 to 400 watts. Most VHF radios in the 70s and 80s were only 10 watts so an amplifier that has gain of 16dB to give you the 300 to 400 watts was quite good.

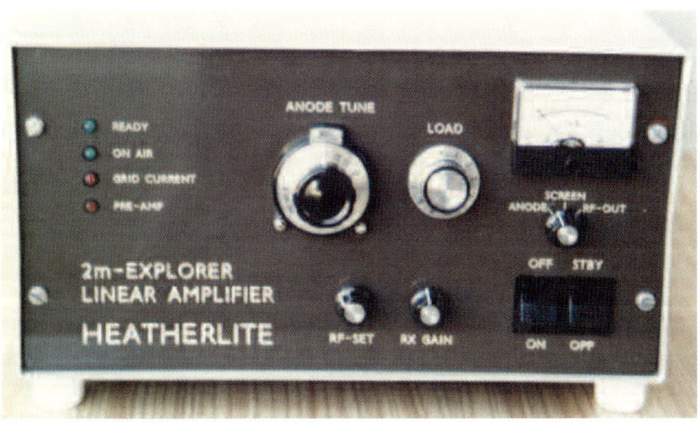

The first 2m 300 watt Explorer

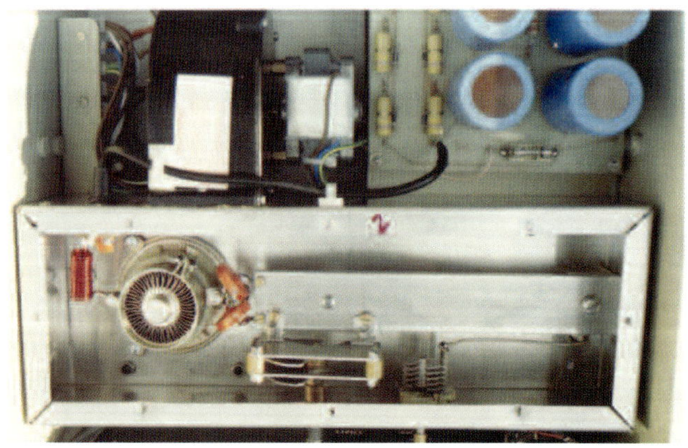

View inside the PA Compartment

My first hand drawn circuit of the 2m Explorer!

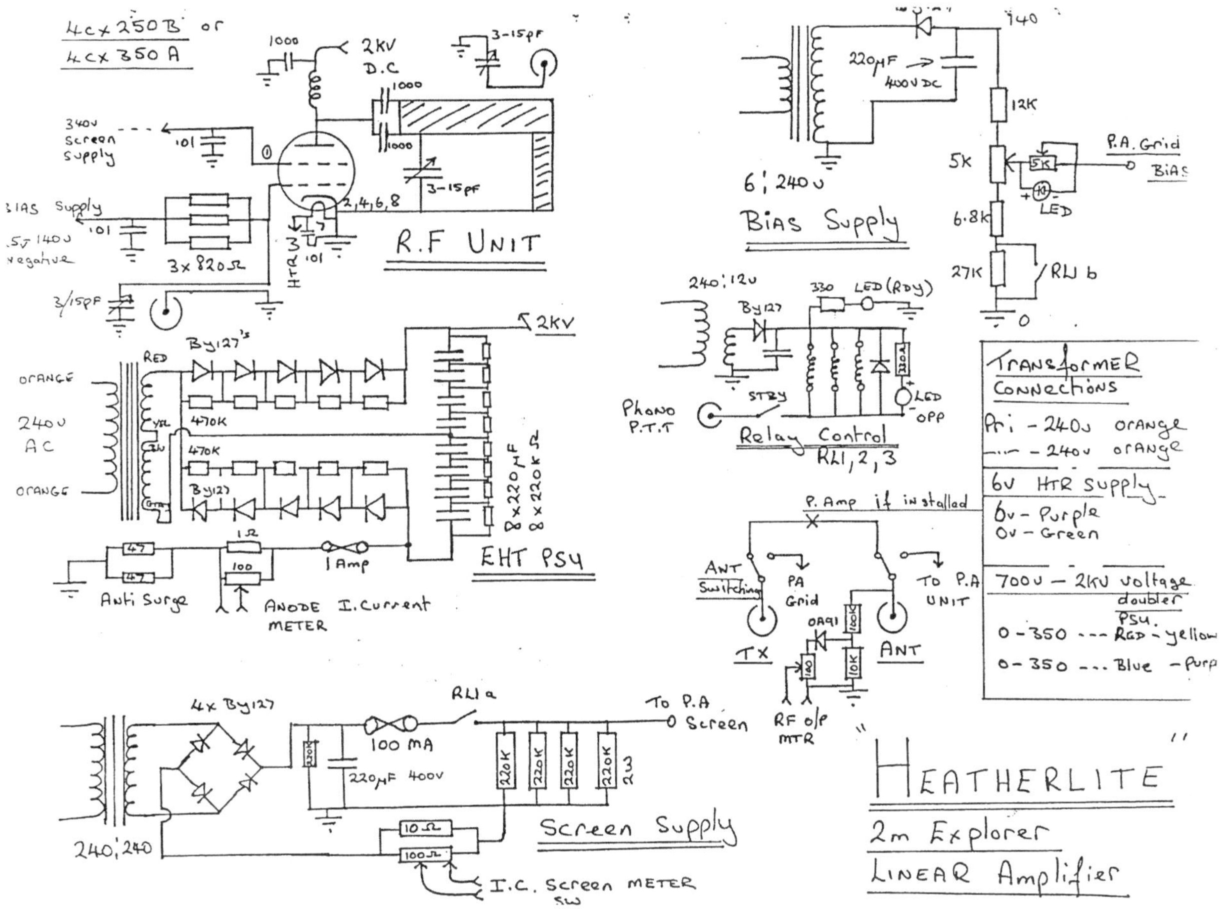

The idea of the 2m Explorer was a combination of articles from the RSGB and ARRL Handbook.

2m Discovery

The first 2m amplifier I made using the 3CX800A7 was after reading articles in the ARRL hand book, so I decided to have a go. The principle was the same using a tuned line in a cavity box though I always preferred the link couple output compared with capacitive coupling. There is a lesser chance of shorting the HT supply to the output. The blocking capacitor isolates the High Voltage and at the current end of the line it is shorted to earth. The input T network gives a very nice 50 ohm input to drive the cathode. Only 25 watts of drive is required to drive the 3CX800A7 to over 1000 watts CW.

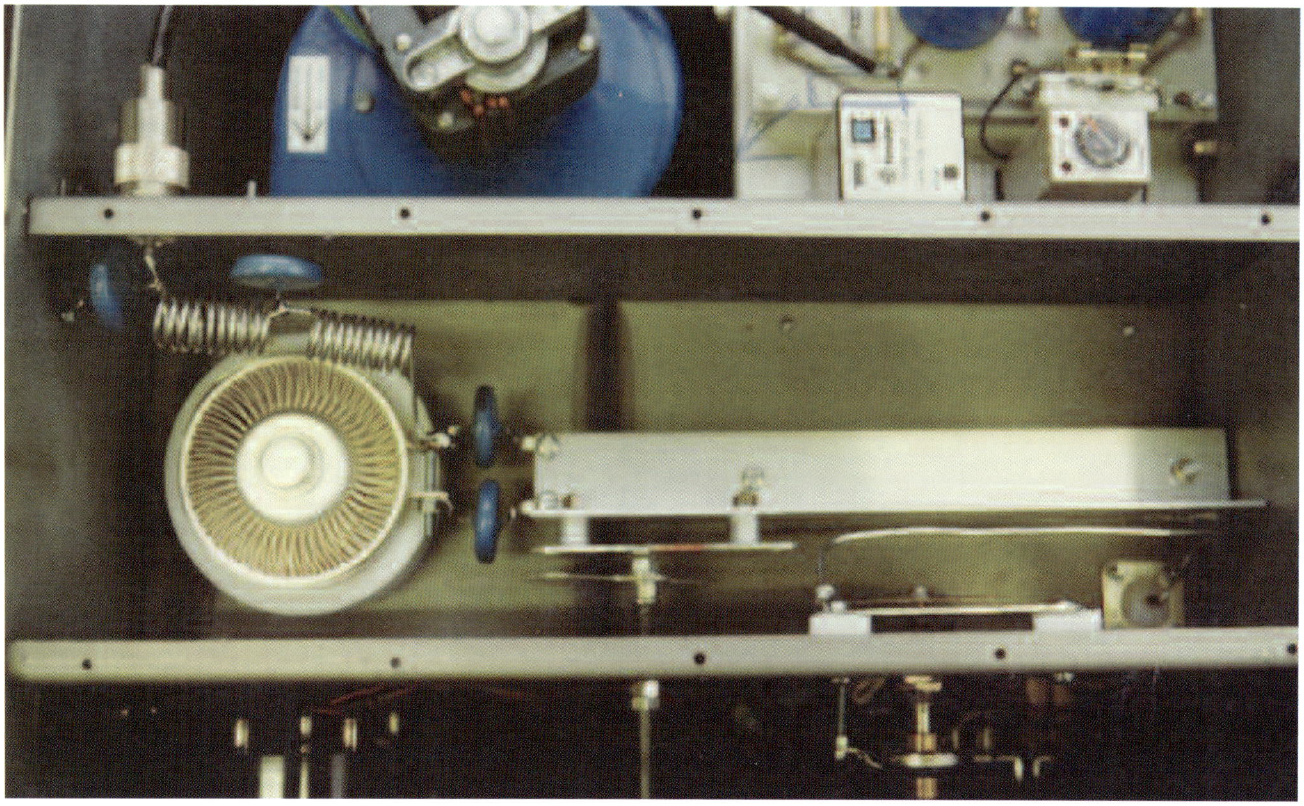

RF deck of the 2m Discovery using a 3CX800A7

6m Discovery Linear Amplifier using the 3CX800A7

When we got the high power licence for the 6m band in the UK, it was a natural progression to produce an amplifier for the band. The 3CX800A7 was a perfect choice because at the time there were plenty of transverters being used with 10 watts or 25 watts output. The single 3CX800A7 with 10 watts will give 400 watts and 25 watts will give over 1000 watts output.

6m Discovery RF compartment

The move to the Russian GS31b and GS35b

I really did not want to move away from the 3CX800A7 as it had become a good friend. It is a super reliable valve with a fantastic gain for a triode. As I have mentioned earlier, the 3CX800A7 and 3CX1500A7 have been my all time favourite valves for 2m to HF. I have built 100s of 3CX amplifiers and even when pushed, they are still very reliable. You should always fit grid protection to limit the grid current to the manufacturers limit, just 60mA for a 3CX800A7and 120 mA for a 3CX1500A7.

Price

When I first purchased the 3CX800A7 the price was around £200 or $300, the price is now $955 plus another $40 for the special valve base, crazy!

At Ham Radio Friedrichshafen around the late 1990s I discovered the Russian triodes. Their price was good and readily available.

I paid about 50 euro for the first GS31 in 1998 well under a quarter of the price of a 3CX800A7 with chimney and base.

GS31b

It has a similar specification to the 3CX800A7 and has a better upper frequency of up to 1000 MHz compared with 350MHz for a 3CX800A.

The gain is not as good and requires a bit more drive. Now with most modern 2m radios delivering 50 - 100 watts the drive required is not problem.

3CX800A7

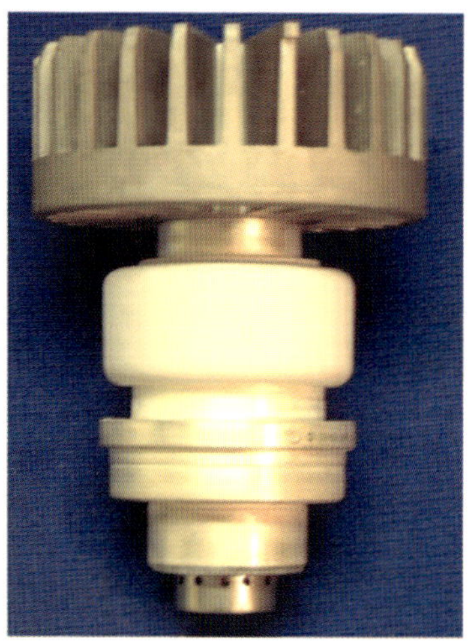

GS31b

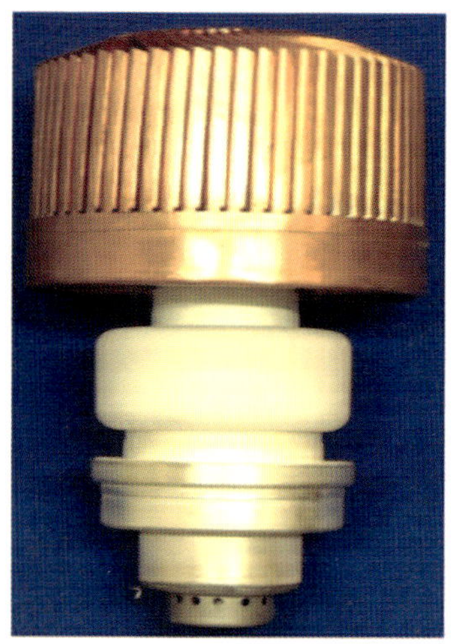

GS35b

A Compact 2m Discovery, using a GS31b

This has been one of my most popular VHF Amplifiers. The GS31b will deliver a cool 1200 watts on 2metres, enough for any serious EME or long range DXer.

2m GS31b Discovery Linear amplifier

The picture above shows the silver plated tuned line.

The output pick-up loop is in the bottom right corner of the PA box. The transformer is a 1 kVA toroidal transformer. More details later in 'How to build an amplifier.

Specifications

Output power	1200W
Drive	up to 100W
DC voltage	2800V
Gain	13dB nominal
Power requirements	230V AC at 10A, 50/60Hz
Cooling	Forced air cooling
Frequency coverage	144 – 146 MHz
Valve	GS 31b
Harmonics	3rd harmonic: better than –45dB
Relays	Coaxial O/P, short throw I/P
Metering	Plate current, Grid current / RF O/P
Protection	Plate over-current, Grid over-current, Over-drive, Internal valve flash-over, soft start and 3 min. timer
Dimensions	14in wide x 9.75in high x 17in deep 360mm x 250mm x 430mm
Start up timer	3 min
Weight	25kg

6m DISCOVERY (GS35)

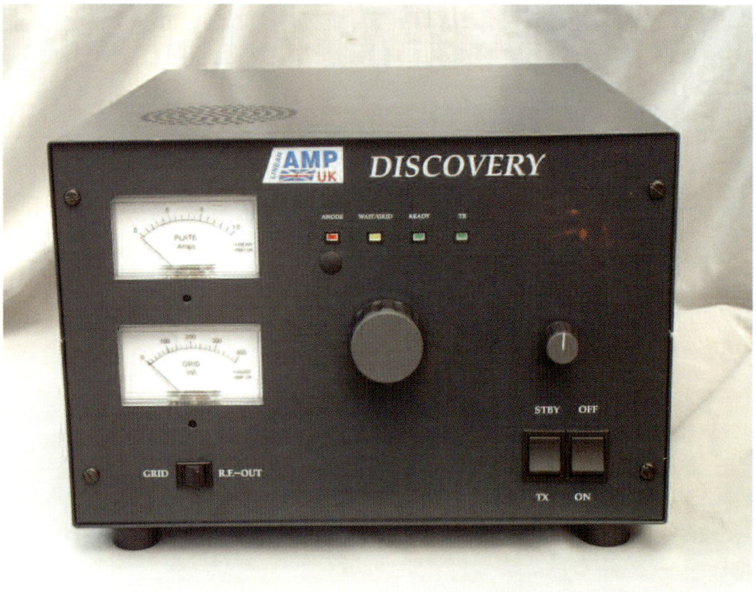

This version of the 6m Discovery uses a Russian GS35b ceramic triode valve. This rugged valve will operate for long periods at an output of 1500W.

The RF compartment is made from aluminium which has been alochromed. The output stage is a Pi- L network with a tuned T input circuit.

A heavy duty toroidal transformer and voltage doubler board produce 3600V. The valve is cooled with forced air from a powerful centrifugal fan.

Protection includes soft start, 3 min start-up timer and anode and grid over-current trip. The relays are sequenced.

Specifications

Output power	1500W
Drive	100W
DC voltage	3600V
Gain	13dB nominal
Power requirements	230V AC at 15A, 50/60 Hz
Cooling	Forced air cooling
Frequency coverage	50 – 54MHz
Valve	GS 35 triode
Harmonics	3rd harmonic: better than –45dB
Relays	coaxial O/P, short throw I/P
Metering	Plate current, Grid current / RF O/P
Protection	Mains circuit breaker Anode and grid over-current protection Soft start and 3 min timer.
Dimensions	14in wide x 9.75in high x 17in deep 360mm x 250mm x 430mm
Weight	30kg

Top inside of the 6m GS35b watt amplifier

The GS35b is the most powerful 6m amplifier that I have designed. It produces 1500 watts with ease. The anode cooler is larger than similar high power triodes like the GU74 and the 3CX1500A7. Due to the size of copper cooler the air pressure going through it does not have to be as great. This model is still made by the DXshop in Wales. The design is stable and the valve only costs around £100 to replace. I have had to change very few GS35b. It is a good reliable valve.

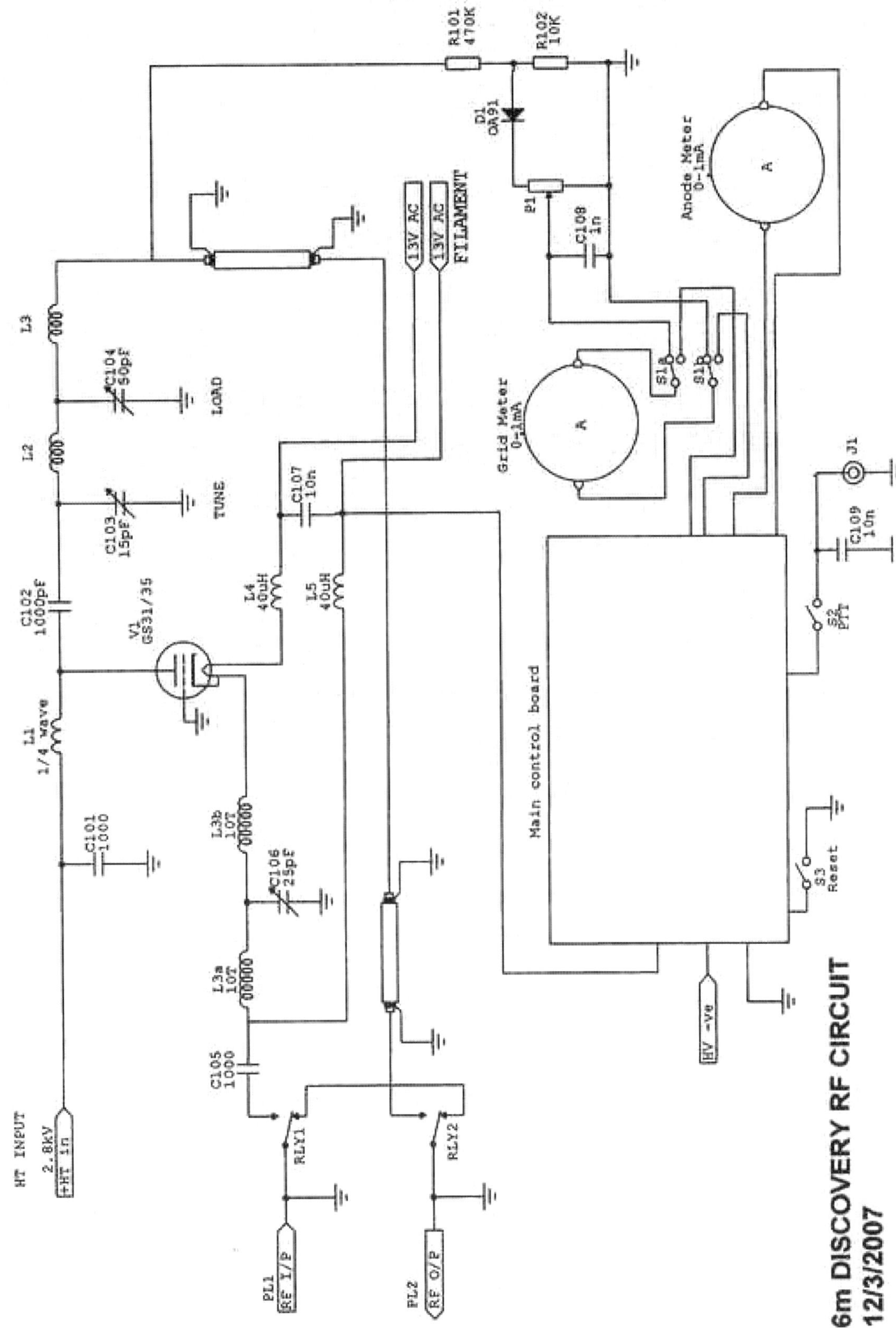

6m Discovery RF circuit ©G3ZRS

Underside of the 6m Discovery

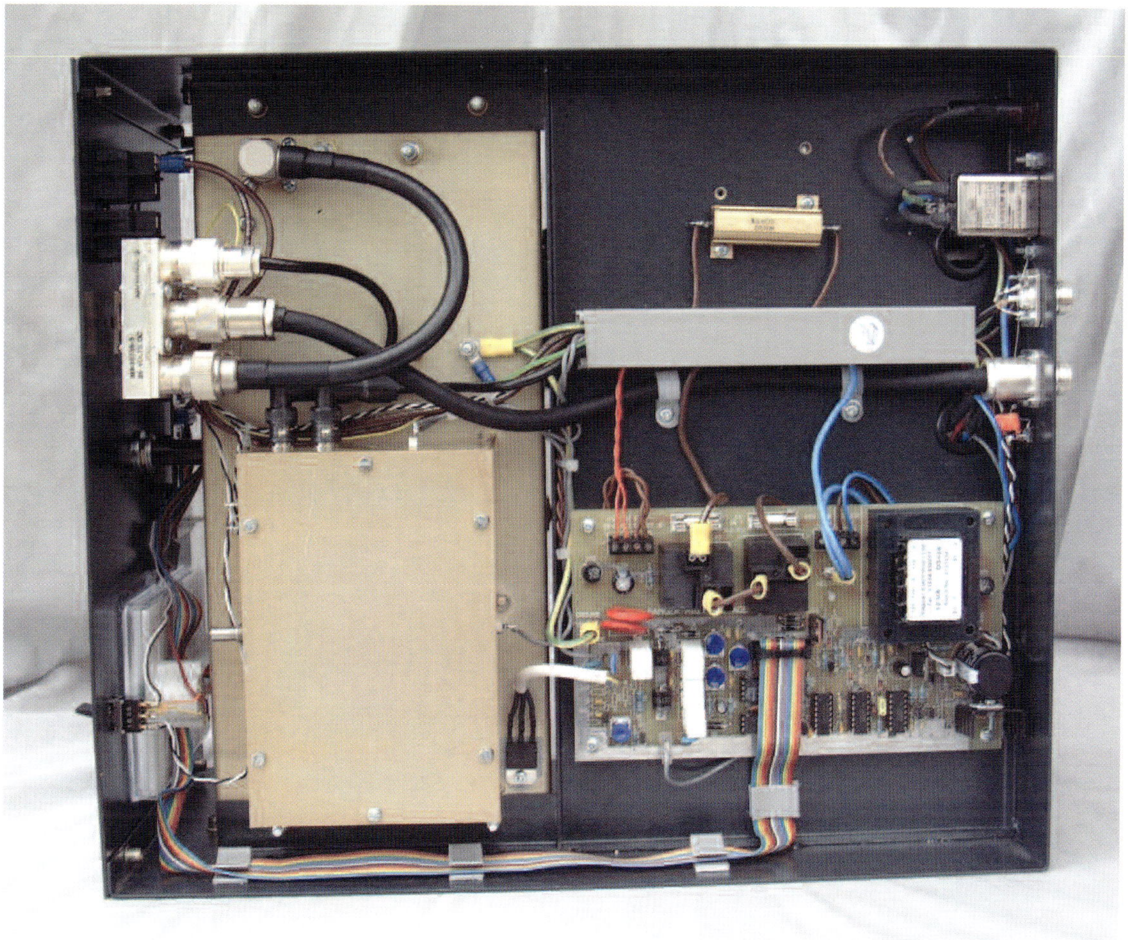

The 6m Discovery showing the underside layout, and this model uses the Ian White GM3SEK control board.

70 cm Discovery using the GS31b

The GS31b is a good tube at 70cm. Do look at valve descriptions! The anode cooler comes off just like all the other Russian Triodes. You can fasten a half wave line between the valve and the anode cooler. The anode valve cooler can be removed by undoing the single large fixing nut which can be seen in the centre of the valve. Eimac tubes needs special finger stock as well the special decoupled base, none of which are required for the GS31b.

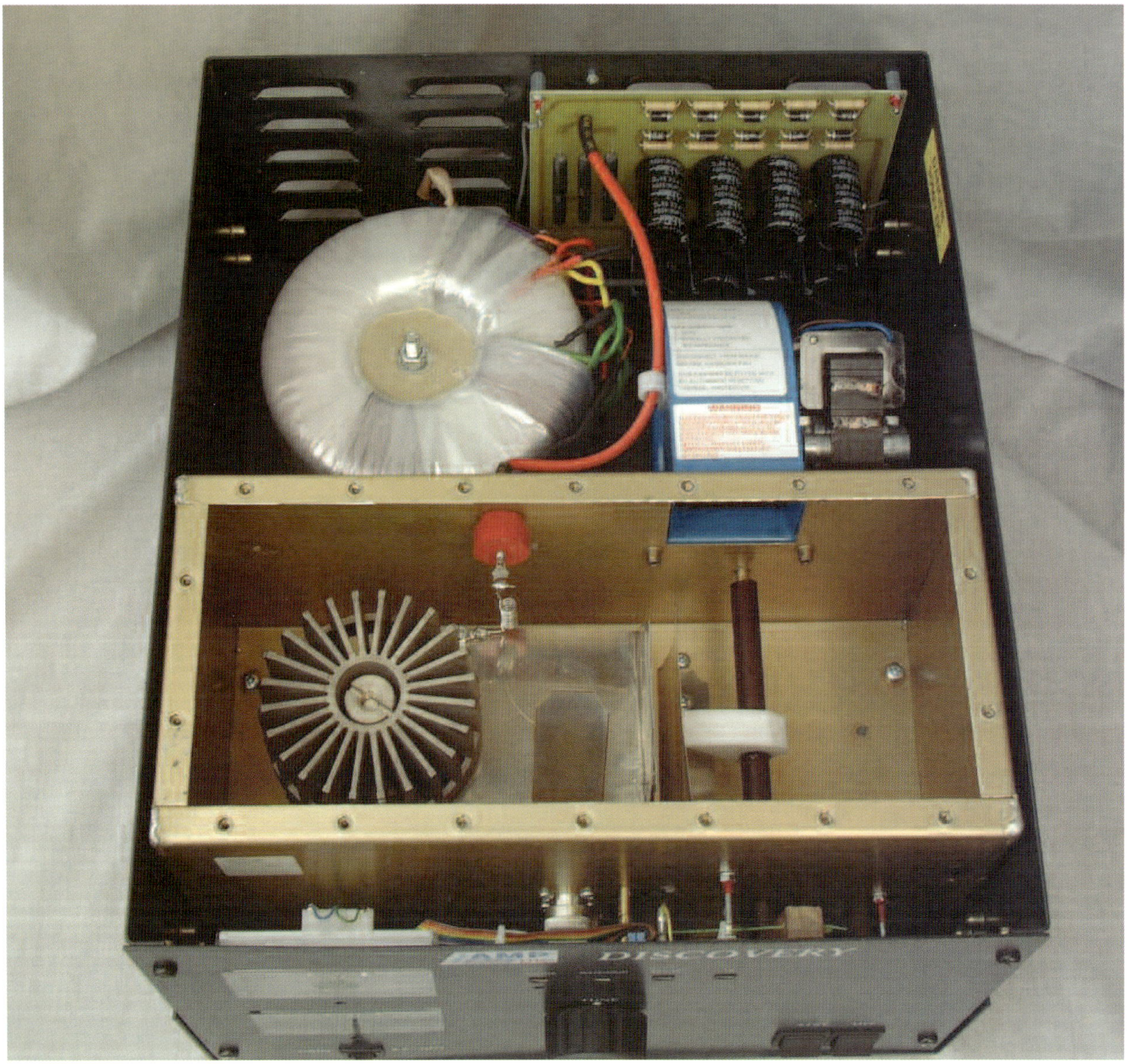

Top inside showing the screened inside cavity box

70cm GS35b 1500 watt

The 70cm GS35b model was only made in limited numbers. I never put it on the market in the UK. All went to Canada and the USA. A lot went to the University of Montreal who were doing research around the 70cm band. Compared with the GS31b model there is much more brass and mechanical bits to make and the cost would be far more expensive.

480,000 miles on 70cm

On the 50th anniversary of Jodrell Bank a 70cm Discovery helped make their first Amateur Radio Earth Moon Earth Contact. The observatory had contacts with Japan, Australia and other countries thoughout the day on the 70cm band.

It didn't take long for me to find that Linear Amp, UK make some 11 dB gain amplifiers for 70cms and so I made a phone call to their managing director, Peter Rodmell, G3ZRS. He was incredibly helpful and immediately put one of his latest 70cm designs into production for me to borrow for the festival. As I collected it, I was able to see the superb internal quality of the amplifiers being built in the workshops at the time.

Prof Phil Diamond
Head of Astronomy Jodrell Bank

A must for all radio hams on 70cm!
Receive on 432.065

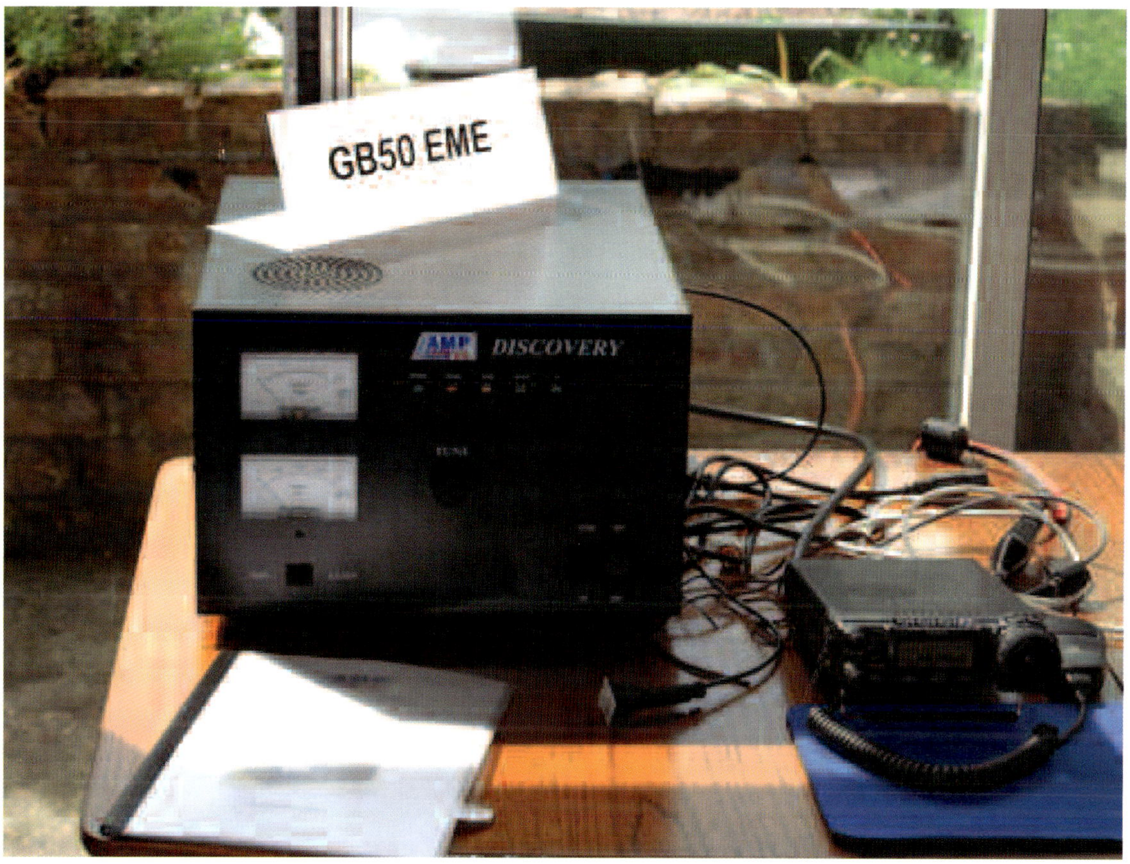

Transmission Frequency 432.065
Using only a IC706 and a 70cm Discovery to work 480,000 miles

GB50 EME UHF transceiver and Linear Amp Discovery 70cm amplifier

CQ CQ GB50EME, Lovell Telescope Manchester, de VK3 UM, Victoria Australia This is Australia calling.

Congratulations on the 50th anniversary of the Lovell Telescope. We wish you many more years of further operation. 73's Doug VK3UM

CONG to celebrate the 50th anniversary of the Lovell Telescope de JA3SGR 73

CQ CQ de GB50EME de SM2CEW Greetings and congratulations to the Lovell Telescope from close to the arctic Circle. 73's Peter.

"GB50EME de KC3RE FM18dp Casanova, VA, USA GB50EME de KC3RE FM18dp Casanova, VA, USA please accept my congratulations to the Lovell Telescope on its 50th anniversary." 73 Martin

Morning Receiving Schedule

07:30 VK3UM Glenburn, Victoria, Australia

08:04 - nice signal and a bit closer to my reading speed - I am very out of practice! I will listen on SSB 08:07 Got you on SSB - copied you very well to GB50 EME Lovell Telescope Manchester from VK3UM North of Melbourne. 08:15 Another station JA? also in the band. 08:00 VK3UM Glenburn, Victoria, Australia

08:30 VK3UM Glenburn, Victoria, Australia

Well copied on CW and voice - thanks Doug!! 09:00 JA3SGR Japan

09:06 week. But largely copied - a colleague from our radio club has arrived and is far better than I. Thank you for the congratulations Toyoshige, we will listen again at 11:00 UT. 11:00 JA3SGR Japan

11:06 As above, 13:30 KC3RE Virginia USA. (In festival EME demonstrations)

The University of Manchester
Jodrell Bank
Observatory

MANCHESTER
1824

Jodrell Bank Observatory
The University of Manchester
Macclesfield
Cheshire
SK11 9DL

tel 01477 571 321
fax 01477 571 618
www.manchester.ac.uk/jodrellbank

19th June 2007

Dear Peter,

Could I take this opportunity of thanking you and your staff for the assistance you gave us this last weekend. The amplifier that you specially built and loaned Ian for his station GB50 EME worked superbly well and his moon bounced signal was surprisingly strong. The initial demonstration of EME with his station as we brought the beam of the Lovell telescope onto the Moon gave the afternoons events an excellent start!

In all, the event was a great success, and we really do appreciate your efforts on our behalf.

We hope that the picture of the Lovell Telescope that Ian has produced may act as a reminder of what was, for us, a really special occasion.

Our thanks again.

Best regards,

Professor Phil Diamond

Head of Astronomy and Astrophysics and Director, Jodrell Bank Observatory

Chapter 10
How to design and make VHF amplifiers

How to design and make 2m amplifiers
How to design and make 4 and 6m amplifiers
How to make 70 cm amplifiers

Making an Amplifier for VHF

Introduction

Making a VHF amplifier has never been easier. In the 1960s to the 1980s there were only a few valves available for radio hams that worked at VHF that would deliver the legal limit of 400 watts. In the 1960s and 1970s the most popular medium power tetrode available to hams was the 4CX250B. The valve was even mentioned by Tony Hancock in one of his comedy scenes of the Radio Ham. It was used for everything VHF and UHF and this followed on from the earlier 4X150 which was introduced by Eimac in 1948. A 4CX250B can be used for 6m, 2m or 70 cm to give around 300 watts output and a pair around 600 watts.

Transceivers in the 70s and 80s were normally only 10 to 25 watts output, so to get to 400 watts a grid driven tetrode was the only choice. Readily available at a modest cost was 4CX250B.

Picture of the 4CX250B

The arrival of the 3CX series of Triodes

In the early 1970s Eimac introduced the 3CX series of externally cooled triodes. Looking similar in shape to the 4CX250B they include the 3CX400A7, 3CX800A7 and 3CX1500A7. These valves are a perfect choice for powers up to 1500 watts. They are much simpler than a tetrode, which require three separate DC supplies. They have good stability with a simple base and do not need a special decoupling ring except for 70cm.

Eimac 3CX800A7

My first 2m /6m Discovery was based on the 3CX800A7. It would give 1000 watts with only 25 watts of drive, a gain of 16dB which was not dreamed of in a triode up to this time with good stability. By using a simple T match input circuit and a tuned strip line you have a good high power VHF amplifier.

The first Eimac 3CX800A7 cost me around $200 each. Now an Eimac 3CX800A7 will cost an astonishing $995 and a Chinese version $465. It is such a shame that the 3CX800A7 has been squeezed out of the amateur scene due to very high pricing from Eimac. It is a fantastic valve for amateur radio. Almost all manufacturers now either use Russian or Chinese valves due to this one reason, price. Alpha used the 3CX series throughout all their production but now nobody uses Eimac tubes in Amateur radio to my knowledge. I am sure William Eital W6UF and Jack McCullough W6CHE who formed Eimac would think this is a very sad day as their first tube for Ham radio was the 150T.

The Russian externally cooled triodes became available at a fraction of the cost and they did not require any special base as the grid ring just gets clamped to earth in the chassis. They are generally slightly larger and do not have the gain of the Eimac equivalent. Fortunately now most VHF transceivers have 50 to 100 watts output so it is not a problem.

Russian Ceramic Triodes for VHF

This is a GS35b which will give 1500 watts on all of HF and VHF frequencies to 430 MHz for a cost of around £100

A guide to Russian valves available: -

GI7b	at 1.8 kV	300 watts O/P with 15 watts drive
GI46b	at 2.0 kV	500 watts O/P with 20 watts drive
GS31b	at 2.8 kV	1200 watts O/P with 80 watts drive
GS35b	at 3.6 kV	1500 + watts O/P with 100 watts drive

A guide to USA / Chinese valves available:-

3CX400A7	at 2.0 kV	525 watts O/P with 20 watts drive
3CX800A7	at 2.2 kV	1000 watts O/P with 25 watts drive
3CX1500A7	at 2.5kV	1500 watts O/P with 60 watts drive

The RF compartments in VHF amplifiers

The RF compartment of a VHF amplifier has to be in two separate boxes, well designed and particularly well screened and decoupled to stop RF leakage. The top compartment on 2m usually has a ¼ wave strip line and on 70cm a tuned half wave line. At 6m, more like HF, a Pi -L network.

The compartment is not critical on size except it has to be large enough to contain the valve with at least a 15mm gap between it and the side of the box. It must be tall enough to make sure you have a good gap between the top of the valve and the top of the box. The consideration is that the valve can be working at voltages around 3000 volts so adequate spacing must be provided.

Another consideration is capacitance of the valve to the side of the case. Any extra capacitance will shorten the line and lower the resonant frequency and lower the Q of the circuit.

The RF enclosure

The most obvious material to use is aluminium, it is easy to work with and very conductive at RF plus being cheap. The thickness does not matter but 1 or 1.5 mm is ideal. The thicker it is the harder it is to work. If you get the box made, get the company to weld the joining corners to make a seal. If not you can use 10 or 15mm aluminium angle and fasten in the inside corners of the box with self tapping screws or nuts and screws.

The high voltage DC should pass through a feedthrough capacitor of 470 to 1000pF, which stops any RF leakage outside of the RF compartment.

It is good practice to supply the heaters using feedthrough capacitors. Though not totally necessary, RF chokes and 1000 pF decoupling capacitors will normally suffice especially on a triode amplifier.

The RF box can look bare! If you get the box made you can ask for it to be anodised which gives the aluminium a nice finish and does not alter screening. A steel RF compartment can be used especially on 6m and 2m but by choice aluminium is always preferable. If you decide to paint the RF box use masking tape to leave the paint off the joining surfaces so you always have metal to metal contact.

The top and bottom covers should be fastened with 4 mm countersunk screws into nut-serts riveted into the aluminium boxes. If not, use coarse self tapping screws that will give you plenty of grip. The fixing screws on the top of the box should be no more than 40mm apart. The bottom box will be ok with a total 4 or 6 fixing screws.

Layout of the underside of a typical VHF amplifier

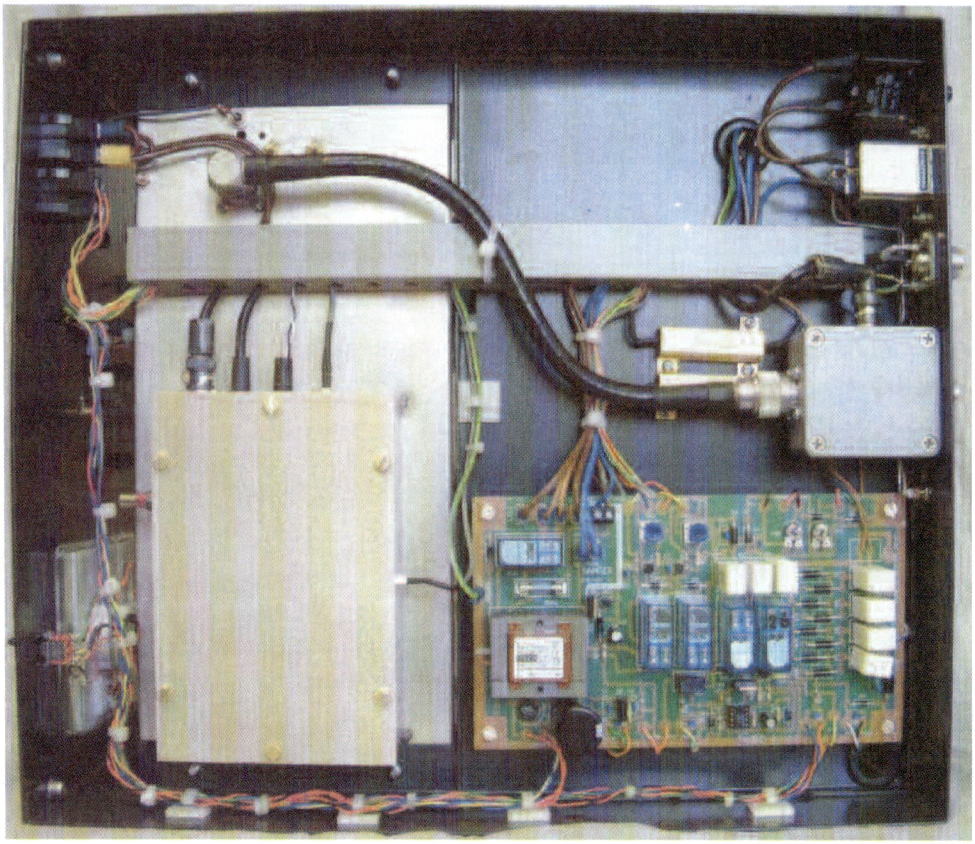

Underside of a GS31b 1200 Watt 2m amplifier

Above shows the G3ZRS/G3TEU control board used in all later Discovery and Challenger amplifiers.

The right hand side is the back and shows from top to bottom, the circuit breaker, IEC connector with internal mains filter, coax in, coax out on the back of the coaxial relay box and then the control board.

The left side is the front and shows from top to bottom, the on/off switch, standby switch, the plate and grid meters with the RF/grid switch selector.

Most of the wiring from front to back is through mini ducting to keep it all tidy.

The input box is 180 mm wide x 45mm high x 120mm deep

Top of a typical VHF amplifier

The picture above shows the RF compartment on the left

The RF box is anodised aluminium to stop discolouring
The 1 kVA toroidal mains transformer, PSU board and snail fan are all on the right
The large red wire carries the 2800 volts from the PSU board to the RF box feed through capacitor.

The outer cabinet is 360mm wide x 250mm high x 430mm deep
The inner RF box is 340mm wide x 160mm high x 160mm deep
The input RF box underneath (see previous page) is 180mm wide x 45mm high x 120mm deep

Note:-most of the components and cabinets are available from Roger Banks at the DX shop in Wales
Tel number 0044(0)1588 620126, he will be pleased to help you.

Start building a 2m Amplifier

Considerations:-

1 Plan where everything is going to go and try not to double stack any components so testing and maintenance can be easily done.
2 Make your RF box first and fit it into the main cabinet, you can use 20mm angle aluminium to sit the box on.
3 Mount the PSU board and fit the transformer last. It is easier to turn the box around and fit all the other components when light.
4 Wire all your AC and DC supplies
5 Make up all the RF leads and fit to the relays and input sockets

Whether you decide to use a GI7b, GS31b, GS35b or a 3CX800A7 the circuit is just the same and the same box will take any of the above valves and the only differences are that the tuned lines have to be made the correct length for each individual valve.

The input circuit is the same for all valves. A standard T match with two coils and one variable 25pF capacitor.

The bias voltage does differ for all the individual valves and the easiest way is to buy the correct zener diode unless you make a variable voltage bias supply. I find the zener route the easiest and most reliable remembering to mount the zener on a suitable heat sink, insulated from earth. You can of course purchase a GM3SEK triode board which will do all the metering and protection as well as give you the Bias supply using a TIP 147.

Cathode bias voltages required for the individual valves

3CX1500A7	=	8 volts
3CX800A7	=	5 volts
GI7b	=	24 volts
GS31b	=	28 volts
GS35b	=	36 volts

It is good practice to have a 200ohm 10 watt resistor across the zener diode. This helps to stop any voltage spikes damaging the zener. It also reduces the amount of current through the zener without affecting the standing current.

I am going to take you through the basics of building a GS31b 2m amplifier. There are not many changes to be made for whichever valve you use.

The anode tuned line is the most mechanical part of the amplifier to make and really is very straight forward.

From the valve it starts with a large DC blocking capacitor rated at least 3x the DC working voltage and big enough to handle the high power of the amplifier. Best is to use a doorknob capacitor to do the DC blocking of 470 to 1000pF. Alternatively there are good wire ended ones rated at 1000pF at 10KV, these are available from Farnell and other suppliers. The doorknob ones come with either 4mm or 5 mm fixing screw unless American and then they have the American type thread.

The Quarter Wave tuned line for 2m

The ¼ wave line is made to be resonant just above the desired frequency and then lowered in frequency across the band by adding capacitance to the high impedance end of the tuned line. You can think of the line the same way as an aerial except the feed is high impedance. The line has high and low impedance spots just the same as an aerial. The low impedance end of the line goes to the case and the high impedance end goes to the valve. A brass flapper can be fastened close to the valve end of the tuned line. The tuning can be done on the valve itself but more often the tuning is done on the line side of the blocking capacitor. The advantage is that if you accidentally ground the flapper to the tuned line there is no DC at this point and is much safer in operation. If you decide to tune directly onto the valve with a plate use a very good insulation dielectric. This can be in the form of Teflon or PTFE sheet and at least 2mm thick.

If you decided to support the end of the line for stability you must use a good quality insulator as the RF voltage is extremely high at this point, use either ceramic, PTFE or Teflon rod.

The output can be a variable capacitor straight off the valve itself at the high impedance end. This is normally fastened directly to an N socket on the side of the case. However the method I have always preferred is to use a link couple at the earth end of the line. This is the low impedance end and with a series capacitor couples the output of the amplifier just like a load control on an HF amplifier. This can be adjusted for optimum output, keeping the grid current to minimum.

There are two variable capacitors in the RF output circuit. C1 resonates the line to the desired frequency, just the same as a tune control on an HF amplifier. C2 is the output coupling and feeds the 50 ohm output coax and normally goes to the RF bypass relay.

Making the line is normally best in brass and about 2 mm thick. If you really want to be posh it can be silver plated at a local silver plating works. We have bought a silver solution that you can rub on yourself and works well.

If you wish there is nothing wrong with choosing aluminium to make the tuned line. The losses are negligible at our power levels. It is more important to use silverplated brass when the powers exceed more than 5kW. Remember RF travels on the surface of the conductor and this is why conductors are often silver plated. Brass or copper goes dulls after a period of time and the silver keeps it looking nice.

If you polish the brass or copper and then spray with a varnish, it will stop the discolouring and keep it looking good.

The line can all be made in one piece by bending it to the desired shape and size. Alternatively you can use straight pieces of metal. Use a U shaped piece of aluminium channel with one side to earth and the other to the tuned line. This method is often easier to adjust if the line is either HF or LF of the desired frequency. If you cannot find any aluminium channel fasten two pieces of aluminium angle together to make a U shape spacer.

The DC supply is fed to the valve anode via an RF choke. This is normally a quarter wave length of 1mm tinned copper wire wound on a former and fed from the large DC feedthrough capacitor.

Consideration for cooling has to be made. A Teflon or PTFE chimney is easy to make, just roll it like a cardboard toilet roll to the size required. Then either staple it together or even better use 2 x 3 or 4 mm nut and screws to fix it in shape. You need to punch holes in the Teflon/PTFE. This is best done with a small hole punch or if you do not have one use a 3mm drill. Use 4 small L brackets to mount the chimney to the lid of the RF compartment. When the lid is on the chimney should cover the entire valve cooler. This forces all the cooling air up through the anode cooler from the bottom and exhaust out through the top of the cabinet.

2m GS31b amplifier

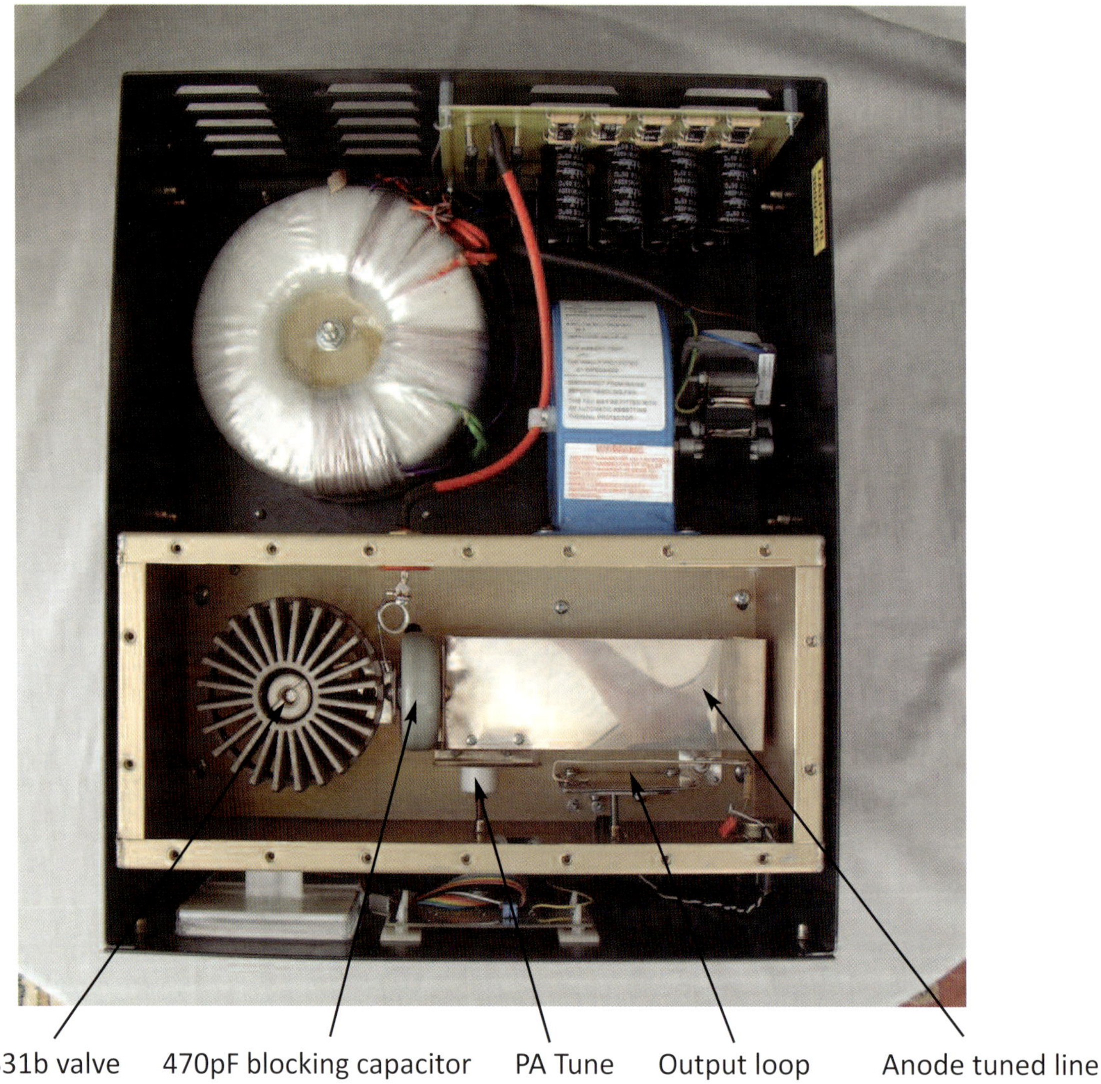

The above 2m amplifier shows the tuned strip line along with the coupling output loop

Between the valve and the anode cooler is an L bracket which fastens the 470pF/20kV blocking capacitor Remember that the anode cooler on most Russian valves can be removed when you undo the centre fixing nut.

The tuned line comes from the right hand side of the blocking capacitor and fastens to the case at the right hand end of the box.

All the inside of the box is pressurised by the blue snail fan. The air then comes up through a PTFE chimney and exhausts through the top of the case.

Bottom centre is the PA tune control. This is a brass L flapper which is fastened to the base of the box. The matching brass L plate is fastened to the valve end of the tuned line.

You will see a solid white PTFE block which insulates the set screw from the brass moving vane. This stops a closed RF loop in the PA which would absorb RF. It also gives a nice smooth tuning of the PA Tune control which comes out on the front panel.

The bottom right corner is the link couple output circuit. This can be a homemade one with 2 plates, one insulated with PTFE and fastened to the end of the pick up wire and the other is a vane that fastens to earth. This is adjusted with a 5mm set screw, PTFE insulation is between the set screw and the moving vane. Alternatively a 25pF to 50pF ceramic commercial capacitor would also work well.

The large red wire carries the 2800 volts from the PSU to the 680pF feedthrough capacitor. This feeds the anode choke and goes to the anode L bracket which supports the 470pF blocking capacitor.

An RF sensing circuit is fastened to the N socket end of the pick up loop. It can be rectified and fed through a potentiometer to give you a RF tuning indicator.

The toroidal transformer is a 900 volt 1 kVA feeding a voltage doubler power supply in the top right hand corner.

GS31b tuned Line

RFC is 7 turns on a 12.5 mm former
The o/p link couple is 7in/175mm

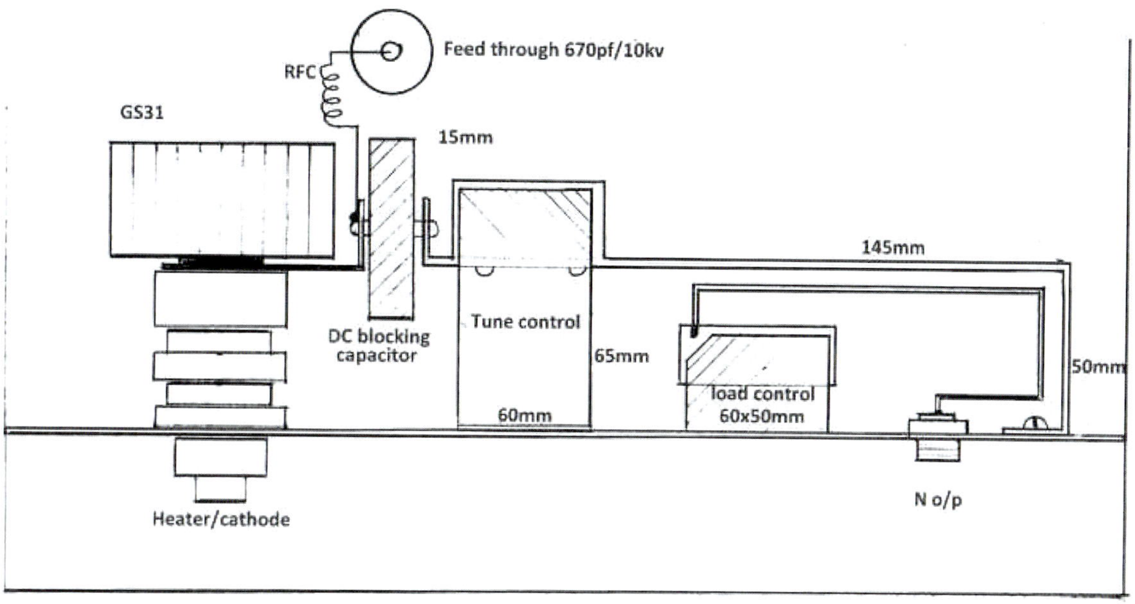

GS31m RF compartment ©G3ZRS

The length of the tuned line is :- 15mm + 145mm + 50mm + 15mm a total of 225mm

The Basic circuit of the GS31b Amplifier

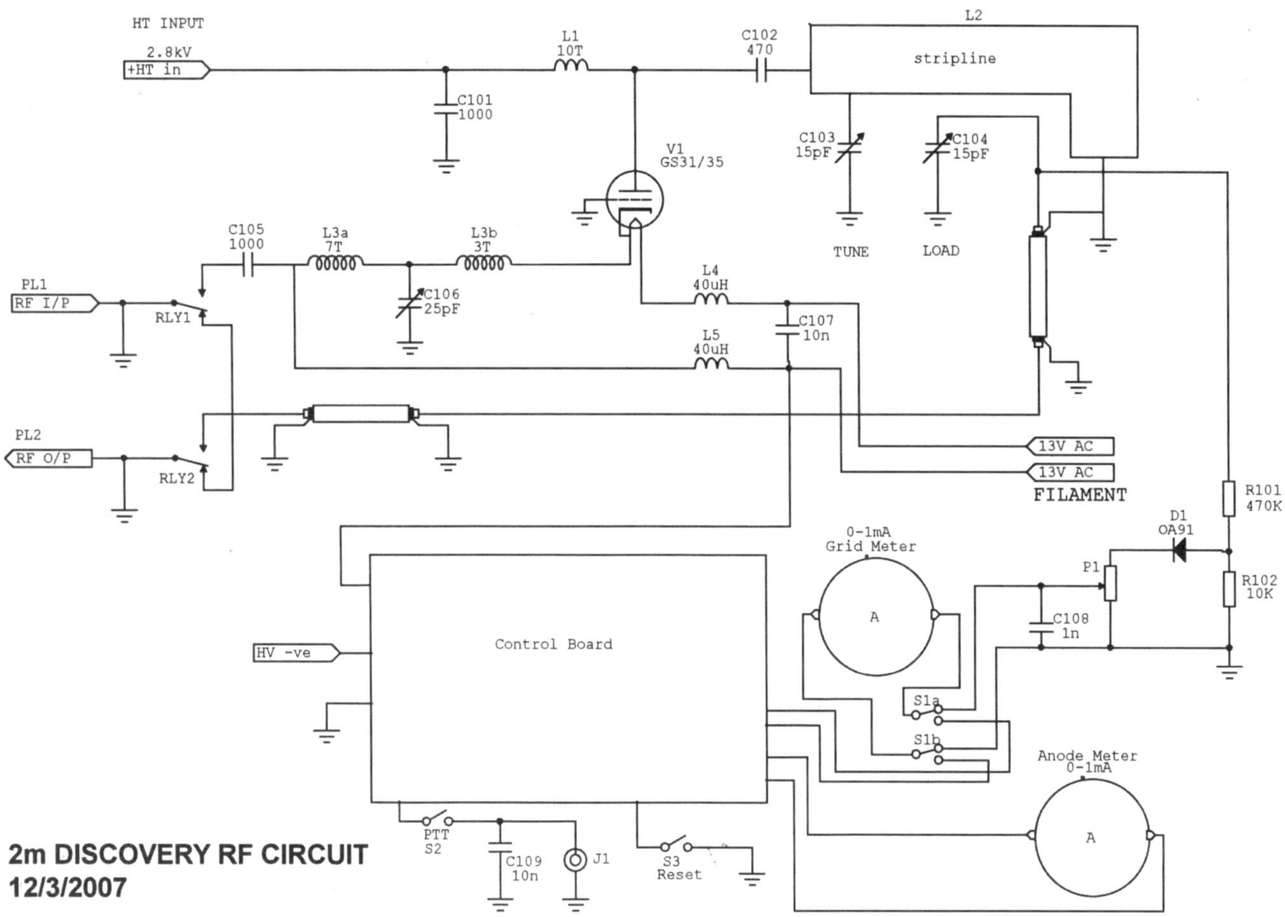

The circuit shows all the RF connections and the control board.

The control board can be as simple or as complicated as you wish. I would recommend the Ian White GM3SEK control board.

The GS31 and GS35 use either a 28 or 36 volt bias voltage.

RF input box of the GS31b/GS35b

The input coil is 10 turns on a 15 mm former tapped at 4 turns from the valve

The GS31b with the heater choke going to the centre connection.

The 10 turn coil forms part of the T match with the tap at 4 turns from the valve. The circuit is two coils and one variable capacitor. In practice it is easier to wind one coil and fasten the capacitor halfway along the coil as shown above. The tap is on turn 4 of the 10 turn coil.

If you do not get a good SWR dip. Try moving the capacitor tap 1 turn either way.

The input coax goes to an Omron single pole relay. It is wrapped with silver foil and bonded to earth This gives a good through SWR without the expense of a true coaxial relay. Of course if you have a coaxial relay at hand by all means do use it.

The valve centre connection can be easily soldered, the next one which is the heater /cathode cannot and a clamp must be made from stainless steel banding or brass.

Valves from time to time can have an internal flashover and will not suffer any serious damage. This is called a 'glitch' and is where your grid protection circuit comes in and de-keys the amplifier. Often if a valve has been stored for a long time the vacuum is not 100%. Just running the heater for a couple of hours will recreate the vacuum.

The GS35b 2m Amplifier

The GS35b is the higher power version of the GS31b and is almost identical except the cooler. The GS35b will take 3600 volts compared with the 2800 volts of the GS31b.

The GS35b will give over 1500 watts with a drive of 100 watts. It does not need a valve base as the grid ring can be just clamped into the bottom of the box with small L brackets made of 2mm brass, 10 mm wide. The cooler is a much larger copper cooler. The heat stored in the cooler means that the back pressure can be moderate compared with valves with a smaller cooler like the 3CX800A7. One advantage is that that the cooler comes off in the same way as the GS31b. The centre is like a cone rather than a machined flat surface. This means you have to make a clamp to go around the valve cooler made of either copper or brass. The large DC blocking capacitor then can be mounted to it between the clamp and the end of the strip line.

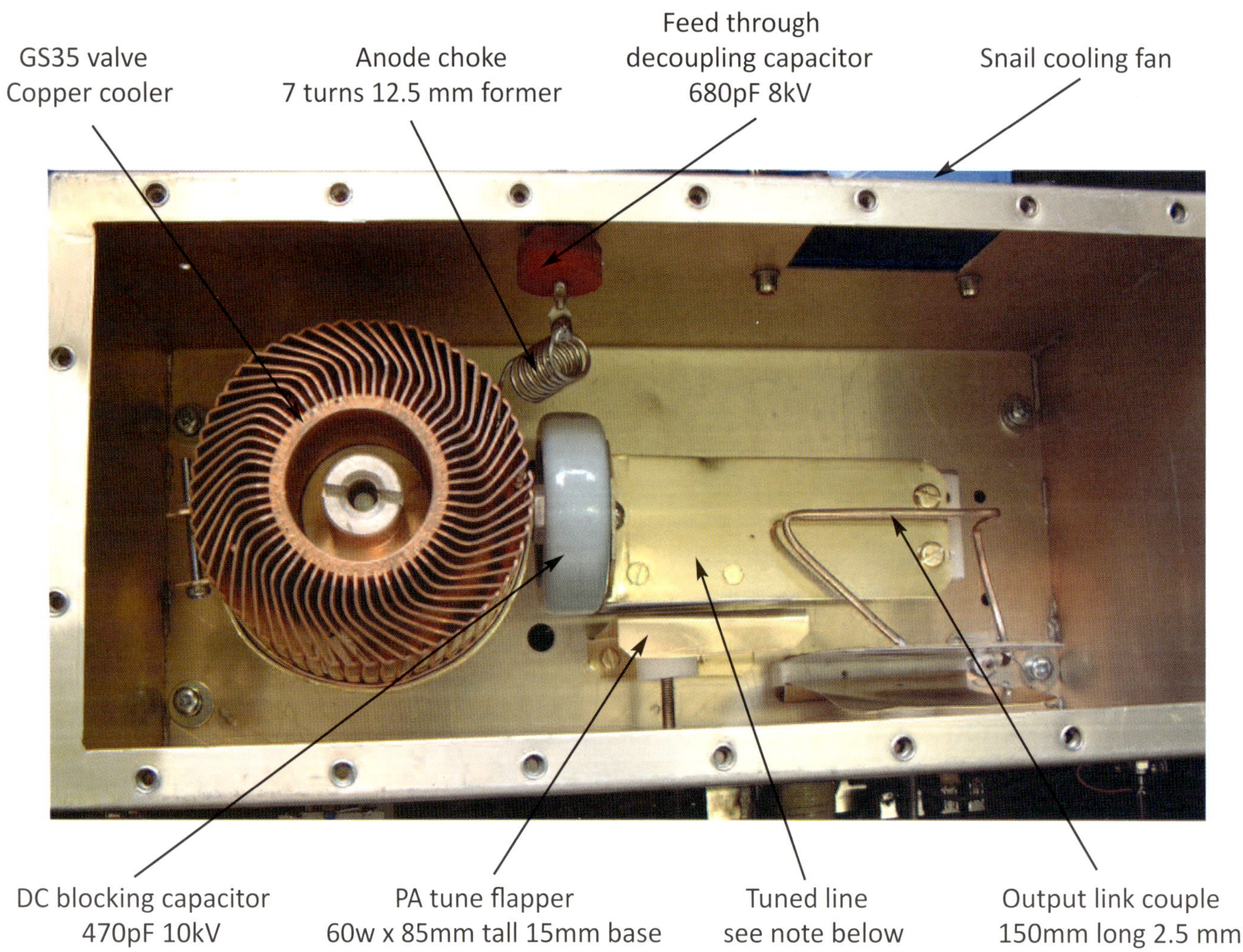

Tuned line = 110 long x 50 wide 50mm high (20mm base to fasten to the bottom of the box)

A total of 195 mm including the 20 mm vertical part fastened to the blocking capacitor

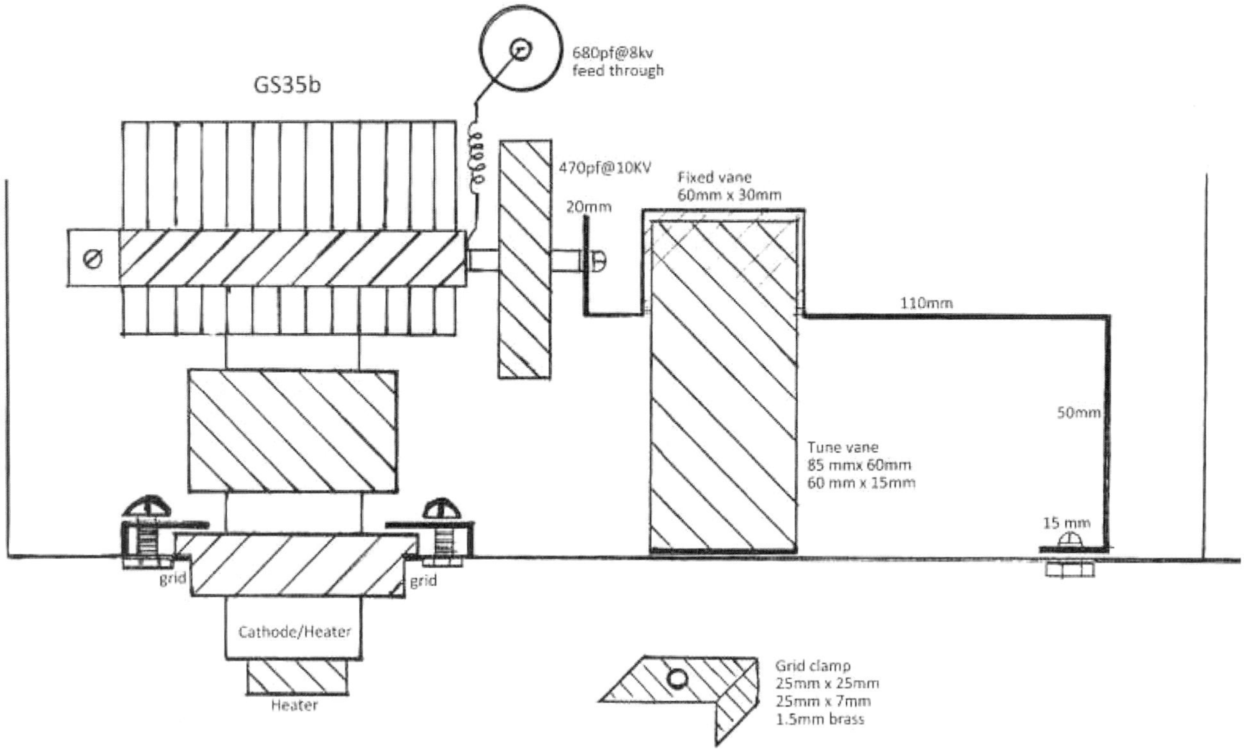

Tuned line in a 2m GS35 Amplifier

The brass tuned line is attached to the anode of the valve via a 470 pF/ 10kV blocking capacitor. The end of the line is connected to the bottom of the PA box to form a ¼ wave tuned circuit.

The line resonates just above the 2m band with a combination of the internal capacitance of the valve and the C from the tune capacitor. It will resonate just above 146 MHz by the closing the tune capacitor it will tune the circuit across the entire 2m band.

The Basic circuit of the GS35b Amplifier

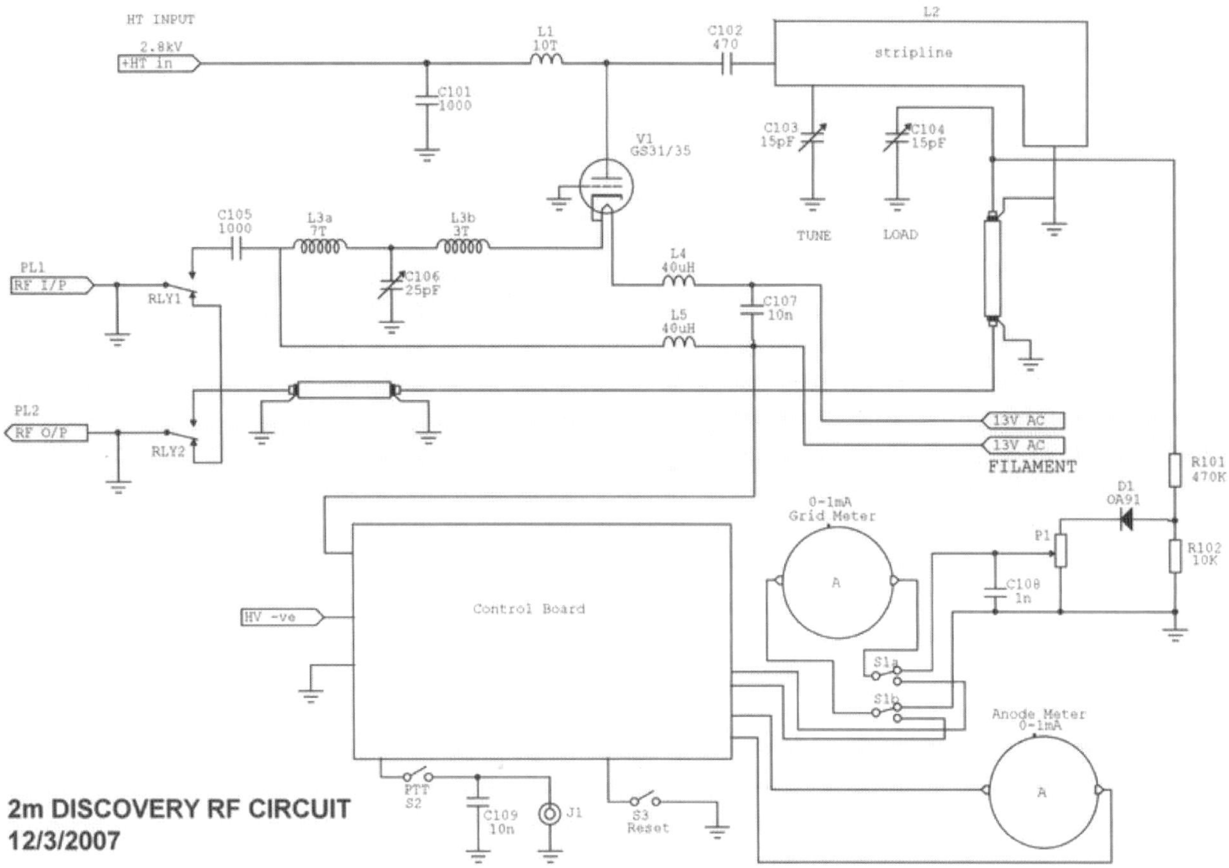

2m Linear Amplifier using the 3CX800A7

My first 2m 1kW amplifier is based on the 3CX800A7, a fantastic valve, having a gain of 16 dB. This is going towards the same gain as a tetrode.

Only 10 watts drive will give you 400 watts out and 25 watts will give you a full 1 kW!

The HT required is low at only 2200 VDC at about 800 mA. You must protect the grid as this is the most sensitive part of the valve. The grid trip circuit must be set to 60 mA, in practice when fully loaded at 1 kW you will not exceed more than 20 to 30mA. The 3CX800A7 is very sensitive to its loading. When properly loaded the grid current is low, but when it isn't the grid current soars high and must be corrected.

The heater is only 13.5V at 1.5 amps which is only 20 watts. This is amazing for a 1000 watt output valve. Of course it is an indirectly heated so you must give it a 3 min warm up time.

There are two more versions of the 3CX800A7 the 3CX800U7 for UHF which has an upper frequency limit of 1000 MHz and is ideal for 70cm. The 3CPX800A7 has a strengthened cathode for pulse applications for NMR (nuclear magnetic resonance) of powers up to 8000 watts. If you find any of these versions as good pull-outs they would make a very good 2m 1 kW amplifier.

The input circuit is a simple T network and the output is a tuned line with a link couple at the current end of the tuned line.

The valve base is a 11 pin special made by Johnson. The valve has a grid decoupling ring and if you manage to find a base that has one do use it. In practice for 2m I find that if you earth the grid pins 4, 7 and 11 well you will not need to use the decoupling ring.

The bias for the valve is only 5.1 volts which can be done with a single stud zener like a BZY93 or a 1N3996. A cheaper alternative is to use 4 wire ended 1N5338 with a 4R7 series resistor like the circuit shows. It is a good idea to add a 180 ohm resistor in parallel as shown in the circuit.

Cooling can be done with most snail fans giving around 30 cubic ft/min.

Between the 3CX800A7 and the strip line you will see the DC blocking capacitors. In this case I use wire ended ceramic 1000pF/10kV which are available from Farnell and other electronic stockists. These work fine and are quite cheap. Alternatively if you have a doorknob capacitor in the range of 1000pF/10KV do use it. When I built this amplifier I had difficulty in locating high voltage feedthrough capacitors. I made a double ¼ wave choke and used 2 x 1000pF /10kV decoupling capacitors. You will see the yellow capacitors in the picture. The DC passes through a ¼ wave length of RG58 which decouples any remaining RF.

Alternative to the 3CX800A7, the GI7b or GI46b.....

Though this circuit is designed around the 3CX800A7 the Russian GI7b or even better the GI46b would work as well in this circuit. They are very much cheaper alternatives with a very similar specification. Both valves do not need a valve base and the heater voltage is the same at 13 volts. The anode voltage is also around 2 kV. The GI7b would give you 400 watts and the GI46b around 700 watts. The drive for both valves is around 20 watts. The tuned line length would be the same and the T match input circuit would also work fine.

Russian Alternative for the same circuit

Russian GI7b Triode

Russian GI46b Triode

The picture below shows the Voltage doubler power supply in the top left corner.

The 3CX800A7 and tuned line are on the right. The toroidal mains transformer is in the bottom left.

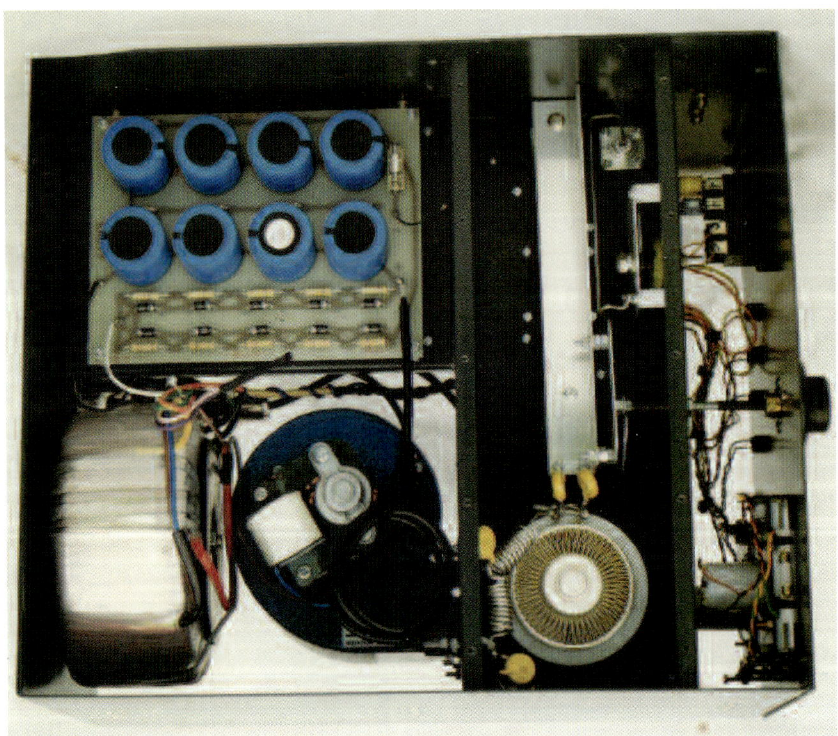

Anode chokes are 2 x 10 turns on a 10 mm former 1 mm silvered wire

All 4 yellow fixed RF capacitors are wire ended 1000 pF / 10kV capacitors

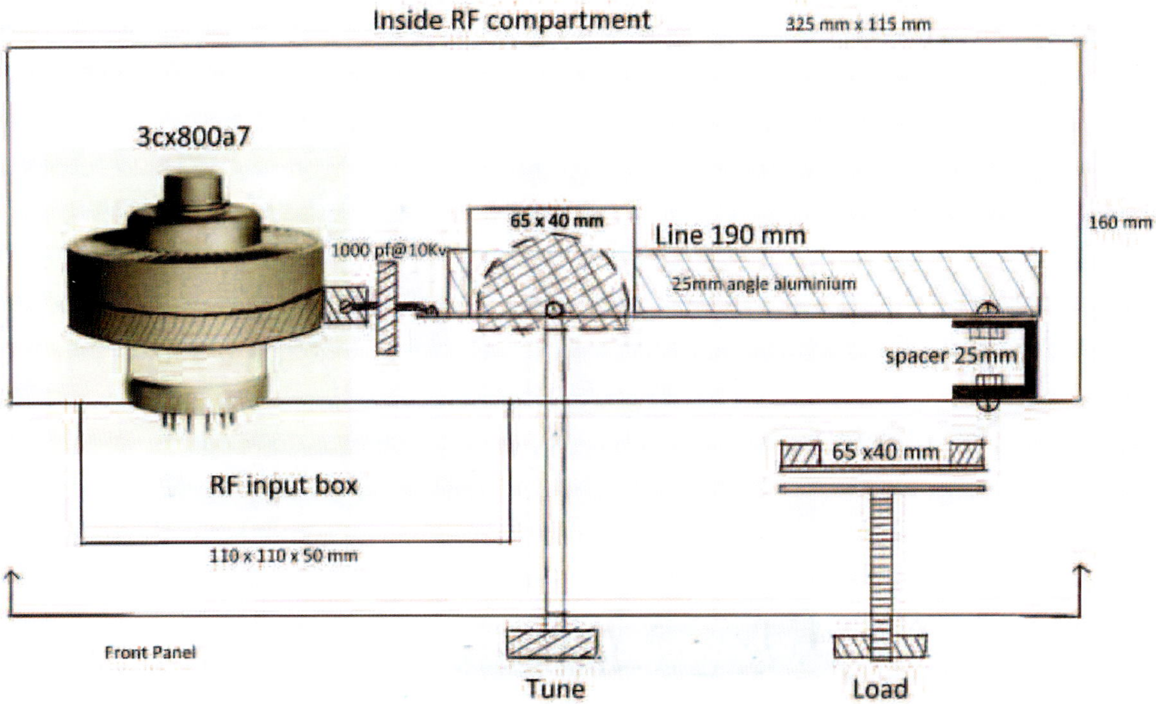

The centre of the valve base is 50mm from the end of the box.

The output coupling link is 200mm long.

PA Tune and Load

You will see by my drawing that the tune and load are both made of a half-moon discs from an old HF capacitor vane. The method is used on both tune and load capacitors. You can use a slow motion dial to indicate the frequency position.

If the circuit is LF or HF you can just adjust the position of the 25mm spacer. The amplifier should resonate at 146MHz with just 30% of the capacitor meshed. It should then tune across the entire band.

The output link goes from the output socket to the load capacitor. It should run parallel with the tuned line as much as possible with a spacing of about 10 mm.

Instead of the half-moon disc for the load control, a 25pF ceramic variable capacitor could also be used. The tune capacitor is at a high impedance, a good spacing of around 6mm is required to stop arcing.

This picture shows the underside of the 2m 3CX800A7 amplifier

The unit makes for a very compact 1kW 2m amplifier.
It is not much larger than an average base station radio.

Front Panel of an early prototype 3CX800A7

In practice once the amplifier is set up the load control never needs adjusting and can be pre-tuned inside and left alone.

RF circuit of the 3CX800A7 Amplifier

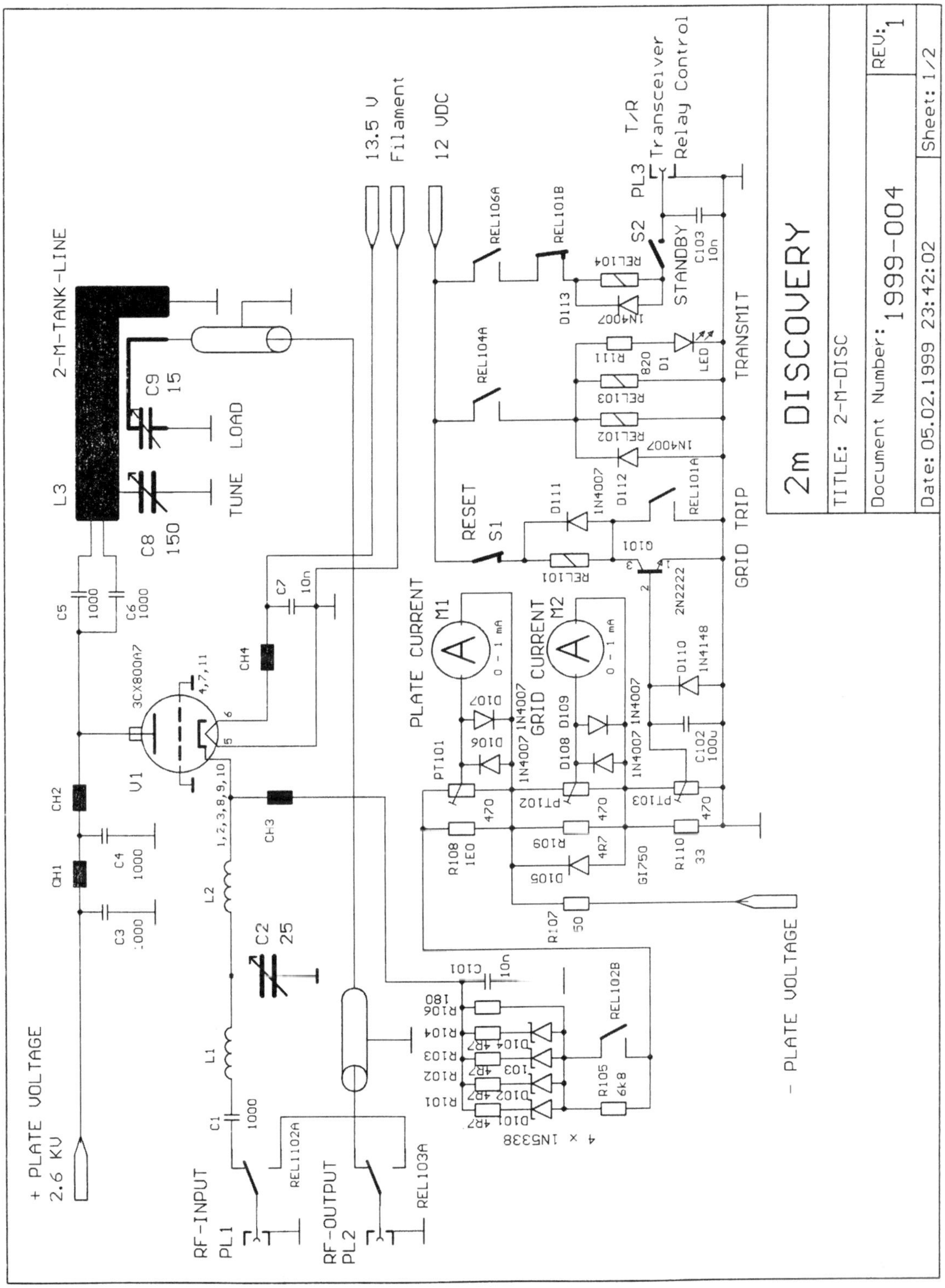

©G3ZRS

3CX800A7 connections :- Anode is the Top cooler Grid Pins 4,7,11 Cathode 1,2,3,8,9,10
Heater pins 5 and 6

Power supply for a 3CX800A7 amplifier

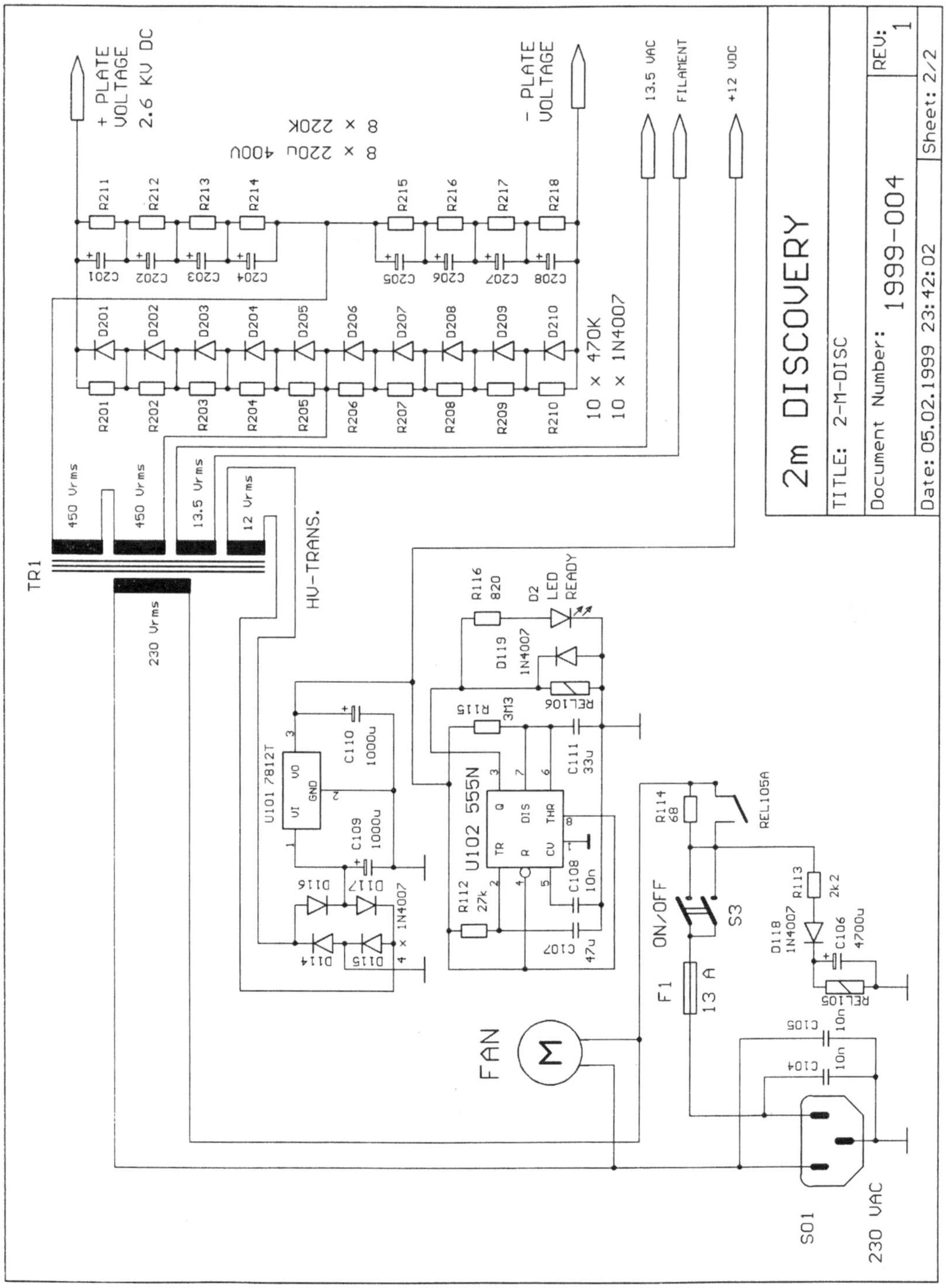

The circuit is a standard voltage doubler with a transformer delivering 900 VDC to give a DC voltage of 2500VDC off load and around 2300 volts DC on load.

The doubler can be made with any of the popular diodes like IN5408 or 1N4007

The start up timer is a simple 555 tmer which drives relay REL 6. This is placed in series with the PTT line so the amplifier cannot be activated until the 3 min have elapsed.

6m and 4m Amplifiers

Triodes on 6 metres

A 3-500Z will work on 6m but now there are better valves to choose from. Ceramic triodes have a higher gain. They are smaller in size and do not cost any more than the old 3-500Z. My first 1kW 6m amplifier used a 3CX800A7. A good pullout with its chimney and special base would make a very nice amplifier and would work well on both 4 and 6m. The 3CX800A7 has a fantastic gain for a triode, in fact only 10 watts drive will give you 400 watt and 25 watts will give you a full 1000 watts. The only problem now is the cost of a new valve, chimney and base make it not worth considering compared with the Russian alternatives.

My next choice would be the Russian GS31b or GS35b. I have used both in my Discovery amplifiers for many years, they are a reasonably priced and work very well.

Both valves have a low internal capacitance which means they work well up to 70cm and are relatively inexpensive. Both valves do not need a valve base as you can fasten the grid ring directly into a hole in the chassis and fastened with just simple homemade L brackets to the case and earth.

Like all ceramic triodes you will need to make a cooling system that directs the cooling air through the anode cooler on the top of the valve. Their gain is very compatible with modern transceivers. With 100 watts of drive a GS31b will give you 1200 watts and a GS35b will give you over 1500 watts.

My choice would be the GS31b having a lower HT supply of 2.8kV and only a slightly lower output. The difference is less than 1 dB to its bigger brother.

All the ceramic valves I have mentioned work on both 4m and 6m by switching the PA coil. All you have to do is short one turn of the PA coil out and you will be very close to 70 MHz. A single T network input will cover both 4m and 6m which make it very easy to change bands. Bring the control knob out to the front panel and mark it either 4m or 6m.

Though the circuitry is VHF they can be basically regarded as HF at 50 MHz when using the ceramic valves. You can be posh and use coaxial relays, but if you are careful with your coaxial leads you can use Omron or similar plastic single or double pole relays. It is good practice to use two separate relays for the input and output which helps eliminate any chance of feedback and taking off. Do also wrap the relays in silver foil and bond to earth. This improves the screening and stops any chance of feedback.

For a power of 500 watts or 800 watts a Russian GI7b/GI46b would be a good choice (look at valve descriptions). They are inexpensive with a good gain and only require 20 watts of drive. No valve base is required as the grid ring can be bolted directly in to a hole in the amplifier chassis. You can make small L brackets to do the job. These two valves would work well as a cheap alternative to the 3CX800A7.

Tetrodes on 6 meters

If you choose a tetrode like a GU74/4CX800 you can drive the radio directly into a 50 ohm load and then feed grid 1 with a 0.1 microfarad capacitor directly into the grid. This will give you a broadband 50 ohm input for both bands without any other tuning for either 4 or 6m. This method will require around 50 watts of drive to give you around 800 watts. The same also applies for larger valves that will give you 1500 watts like the 4CX1000 or GU84 /4CX1600.

Staying with tetrodes for a moment! If you have a low power driver like a transverter or an old radio that gives only 10 - 25 watts output, a tetrode would be good choice. You can make a simple input circuit using a parallel tuned circuit and a link couple as the input. This will give the valve a lot more drive than using a load resistor. You will then get the full power of the valve of around 800 – 1000 watts with only 25 watt drive.

The choice really is almost 100% Russian as their ceramic tetrodes are well proven and the cheapest available. If you find a good Eimac 4CX1000 or 1500 they would make a nice amplifier for 4m/6m, but remember the Chinese also make copies of the 4CX series at a much more affordable price. You will still require a matching bayonet valve base.

Choosing a tetrode requires three DC supplies compared with a single DC supply for a triode and the screen voltage has to be regulated.

GU74 / 4CX800

4CX1000

Using older amplifiers on the 6m band

If you have an older valve amplifier that uses a 3-500Z like the Linear Amp Hunter, Ameritron AL80, Drake L75 etc, it is quite easy to make the amplifier into a single band amplifier for 6m rather than trying to make it HF multi band + 6m.

The above picture is a guide of how a 3-500z amplifier should look for 6m

The picture is of a Linear Amp UK Hunter 6

The main considerations using a commercial 3-500Z HF amplifier on 6m

The **Tune capacitor** has too much capacitance, even at its minimum. The spare vanes must be removed or replaced. It could be stripped and rebuilt removing the moving vanes on both the rotor and stator. You only leave two fixed and two moving vanes as the capacitance required is only 50 pF. All commercial amplifiers that use the same tune capacitor isolate 80 % with a switch and bring it in again for the other lower bands.

Load capacitors in commercial amplifiers are multi section with a capacitance of around 1000 pF. You can sometimes use just one section which would reduce it to around 300 pF maximum, this would be more suitable.

The PA coil details for 6m

The PA coil has to be removed and shortened. The 10m to 20m coil can be shortened and will work fine.

A **single 3-500Z** will require 4 turns of 6mm tube on a 35mm former plus another 5 turns of 3mm tube on a 12mm former to make it into a Pi-L network. This will give better harmonic rejection. (See picture and circuit diagram).

2 x 3-500Z will require 3 turns of 6mm tube on a 35mm former plus 5 turns of 3mm tube on a 12mm former.

An HF Anode choke will normally work fine on 6m. If you do get hot spots replace it or reduce the number of turns to 45 turns of 1mm wire on a 20mm former.

Cathode choke will be fine and will not need any modification

Input Tuned circuit for a single 3-500Z

The input circuit I use is a simple T match which you can see in the picture below. It is easy to make and gives a nice 50 ohm input.

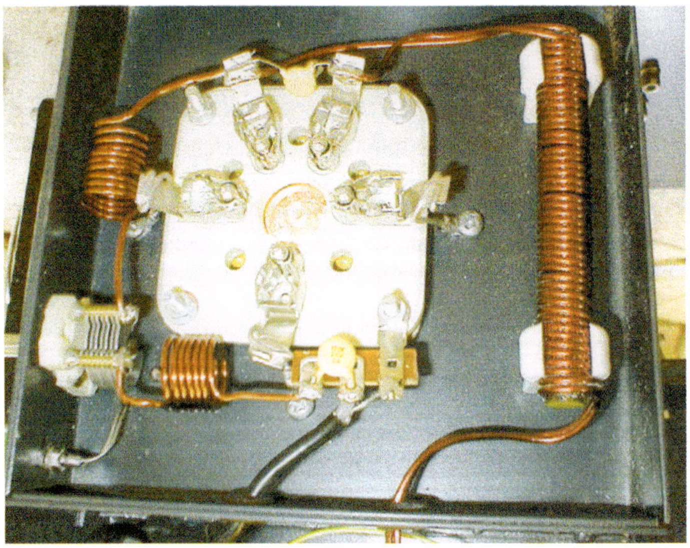

Valve base at the bottom of the 3-500Z 6m amplifier

The input coil (top left) is 10 turns and the lower left one is 7 turns, both on a 16mm former using 1.5 enamelled wire. The cathode choke (right) is 2 x 30 turns on a 6mm ferrite rod and is not at all critical in size. The variable capacitor is 50 pF to handle up to 100 watts.

Isolation

Isolation is a must as any feedback will make the amplifier take off. Most HF amplifiers use a common relay for the RF in and out. I have found it advisable to use separate input and output relays or better still, if you have a spare coaxial relay, use it for the output. If you use an open relay in a plastic case, wrap the case with silver foil and bond it to earth.

If the amplifier is built totally open like most Ameritron amplifiers. It is better to put the input circuit into a small metal box inside the amplifier for better isolation. It could also be mounted on the back panel with the 50pF T-match tuning capacitor available to adjust.

APC (anti parasitic choke)

All HF amplifiers employ an APC fastened to the top of the valve. This acts as a choke for frequencies above 30 MHz and so must be removed. A new one must be made like the one in the picture. Or a better option is a loop made of nichrome steel banding 120 mm long and 6 mm wide with 3 x 150 ohm carbon resistors in parallel to give 50 ohms.

In Operation

A converted or newly built 3-500Z amplifier for 6m makes a very nice project. It should give you around 600 to 800 watts depending on the HT voltage. If you do it with a Heathkit SB220 using 2 x 3-500Z you should get over 1000 watt. From an amplifier that was built over 30 yrs ago it makes a very good recycled amplifier project! Good value for money! Do look on the web as there are other hams that have done it.

If you do a conversion it means you are over 50 % of the way there. You will not have to build the power supplies and the metering and PTT circuits which are already done. You also save time on not having to make the case and drill all the holes for the components. Due to the age, some older amplifiers will require some TLC while it is stripped down. I would upgrade the PSU diodes to 1N5408 and also renew the smoothing capacitors to new ones which are much smaller in size. Use 220 microfarad at 500 volts as this will give you a better margin. While you have the capacitors out, check the parallel discharge resistors with an ohm meter, they are normally 50k or 100k. It is important that each one is close in value to the others. If a capacitor goes open circuit it puts a very large voltage across the discharge resistor and usually takes it out, so it is worth checking. If a faulty resistor is not changed, it will probably take out the newly fitted capacitors in its place.

I have never used a 3-500Z on 70 MHz so will it work or not? The answer is I do not know. The stated upper frequency is 110 MHz so with care it should go to 70 MHz!

RF Deck of a 3-500Z 6m amplifier

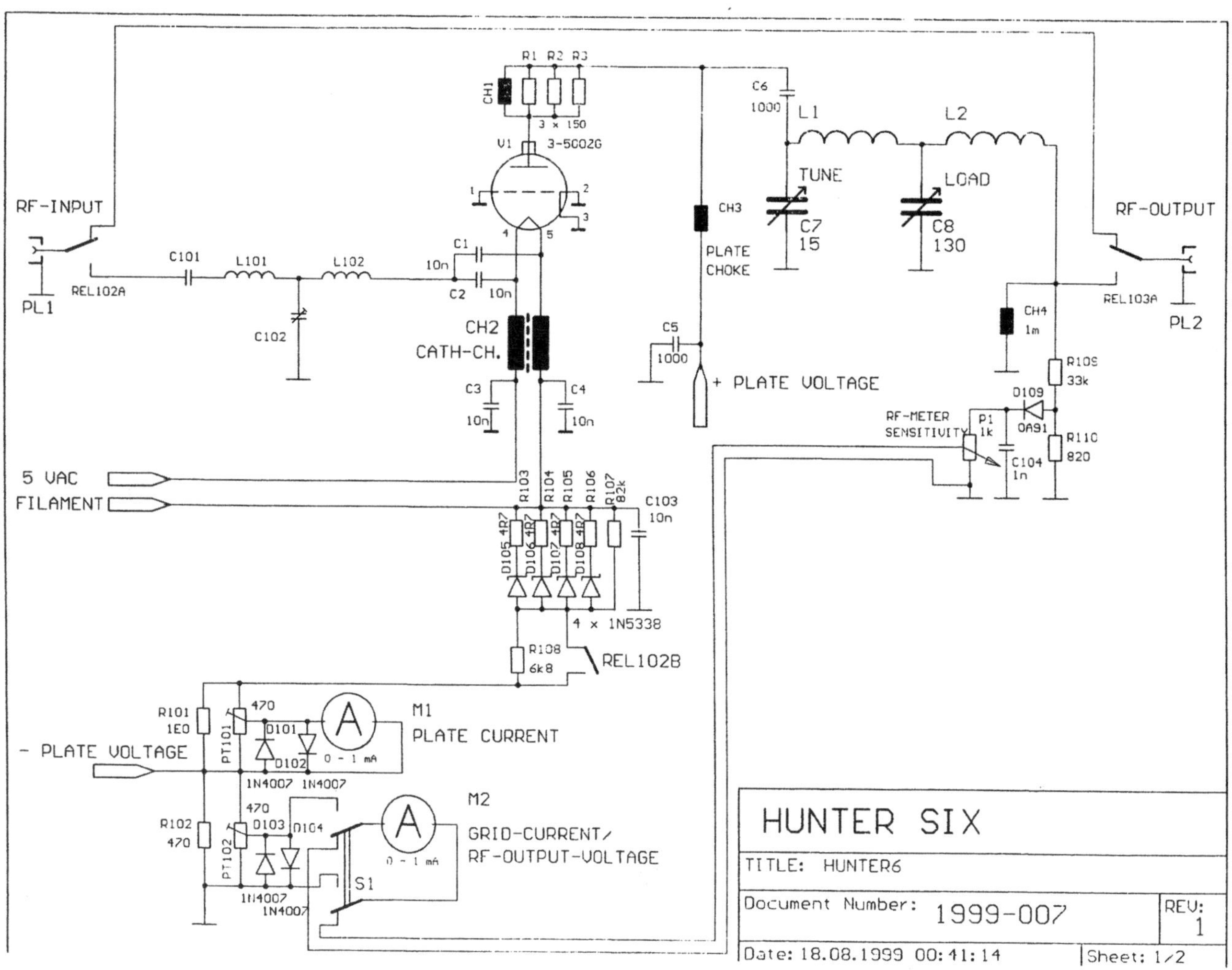

Coil and capacitor details

L1	4 turns 6 mm copper tube on a 35mm former
L2	5 turns of 3mm enamelled copper wire on a 12mm former
L 101	7 turns 1.5 mm enamelled copper wire 15 mm former
L 102	10 turns 1.5 mm enamelled copper wire 15 mm former
C7	15 pF wide spaced
C8	130 pF wide spaced
C102	50 pF ceramic variable

A single 3-500Z is probably the simplest amplifier for 6m you could make. The heater is 5 volts at 15 amps so soft start inrush protection is advisable. The bias is done with 4 x 5 watt zener diodes in parallel. A single BZY 93 would do the job also just as well.

If you are building from scratch I would not buy a 3-500Z. My favourite valve would be the Russian GS31b. It is cheaper and has a much better specification. The top end frequency is above 500 MHz and it will easily give you almost twice the power of a 3-500Z.

Power supply and control circuitry for a 6m 3-500Z amplifier

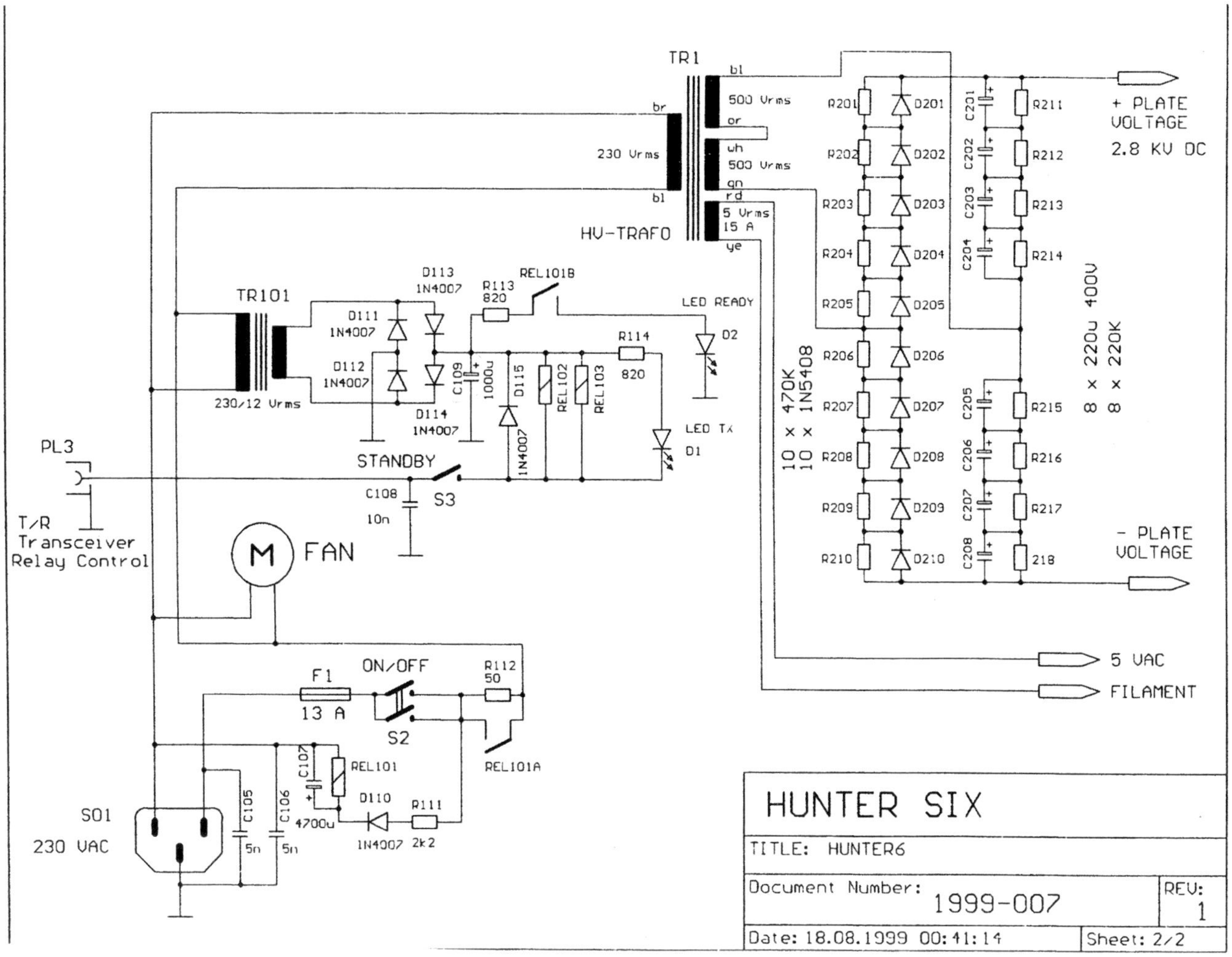

A 3-500Z power supply

This power supply shown will run a 3-500Z amplifier to about 750 watts. From the mains input you will see R112, REL 101A etc which are part of the soft start circuit which is important for the long life of the valve. The circuit is a standard voltage doubler running at 2.8kV. The transformer has 2 x 500 V AC windings in series to give 1000 volts then the doubler takes the voltage up to 2840 DC.

3CX800A7 1000 watt 6m amplifier

The 3CX800A7 is wonderful valve on 6m. It has been one of my favourite valves for many years. It is very small and stable and operates at relatively low voltage of just over 2kV. It has fantastic gain of x 40 some 16 dB. It is fitted with a grid ring that can be used, but not totally necessary on 6m. By just earthing the grid pins on the valve base gives adequate stability for 6m. The 3CX800A7 is ideal for transverters and lower power radios from the 70s and 80s. The valve base is a 11 pin Johnson special, you will also need a chimney which you can make yourself or buy. You can make it using a piece of PTFE tube or sheet. The valve has a delicate grid, so maximum current of 60 mA must not be exceeded. Some amplifiers do not fit a grid trip circuit but I would always recommend you do. While tuning the load control it makes a big difference to the grid current. When it is properly loaded the grid current will be below 20 mA. When off tune it can easily exceed the 60 mA and will easily damage your expensive valve.

The 3CX800A7 has been used for many industrial applications as well as radio ones. Like all things they are getting replaced with solid state alternatives, they are becoming available on eBay and other web sites so do look out for a good pullout. You will have re-mortgage the house or sell the wife to buy a new Eimac 3CX800A7!

A Compact 6m Amplifier using a 3CX800A7

Front panel of the 6 metre 3CX800A7 amplifier

The PA of the 3CX800A7 amplifier, the valve and anode choke are on the left. DC comes in through a SO239 socket. It is then decoupled with the blue 1000pF /10kV capacitor. At the time I did not have access to high voltage feedthrough capacitors which would be a better choice. They are now available from RF parts in California or Roger Banks at the DXshop in Wales.

The tune capacitor is just two rotors and two stator vanes wide spaced, the capacitance is 15 pF which is adequate to tune the amplifier across the band.

Load Capacitor is a 140 pF close spaced

The Pi- L network:-

The PA coils are 2 coils made of the same 3mm copper tube without cutting.

4 Turns of 3mm tube on a 30 mm former
The second half is 5 turns of 3mm on a 12mm former

Anode RF choke is 40 turns of 1mm varnished wire wound on a 20mm PTFE rod

The coupling capacitor is 1000 pF /10kV capacitor. This is fine but if you do have a small doorknob type capacitor it would be physically better and easier to mount.

Below shows the bottom view of the 3CX800A7 base and input circuit

The **cooling fan** blows in from the side of this input box. It is an Airflow Developments 33 BTFL

The picture shows **the input T network** which is 2 coils and 1 capacitor
L1 and L2 are both 10 turns wound with 1.5mm enamelled copper wire
C2 is a 100 pF close spaced variable

The valve base is suspended with 3 legs. It supports the 11 pin Johnson base.
A 65mm hole is large enough for the valve air supply.
The valve chimney is made of a PTFE tube with a 65mm internal diameter. It can also be made with PTFE sheet rolled like a toilet roll.
A proper chimney is available from Eimac distributors and is a SK-1900.

The input relay is a single pole relay mounted in the bottom of the input box. It is wrapped with silver foil and bonded to earth which gives a good through VSWR.

DR 3 and 4 are 5 amp 40µH ferrite chokes.

Circuit diagram of the 3CX800A7 RF deck

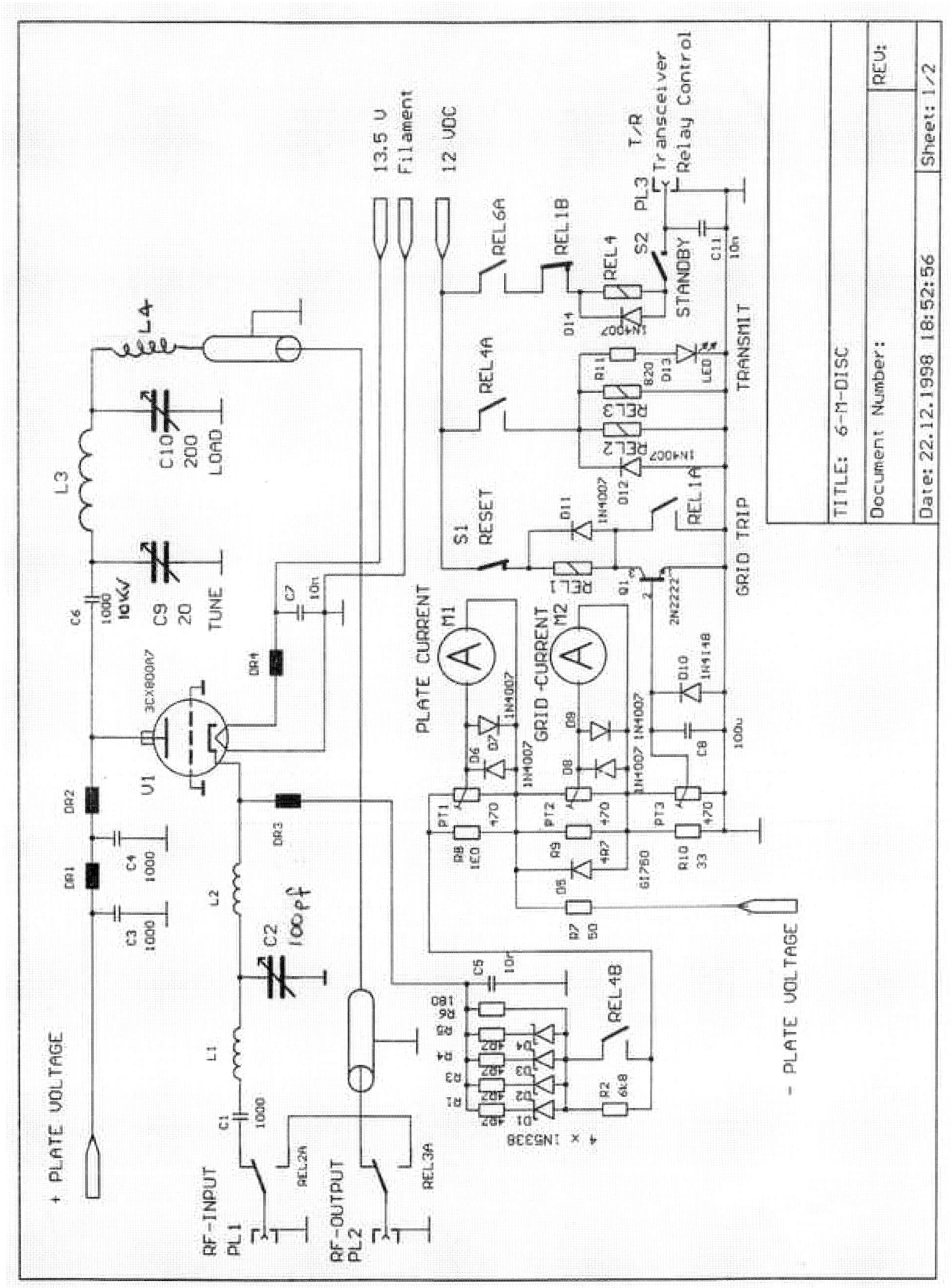

Circuit diagram 3CX800A7 power supply

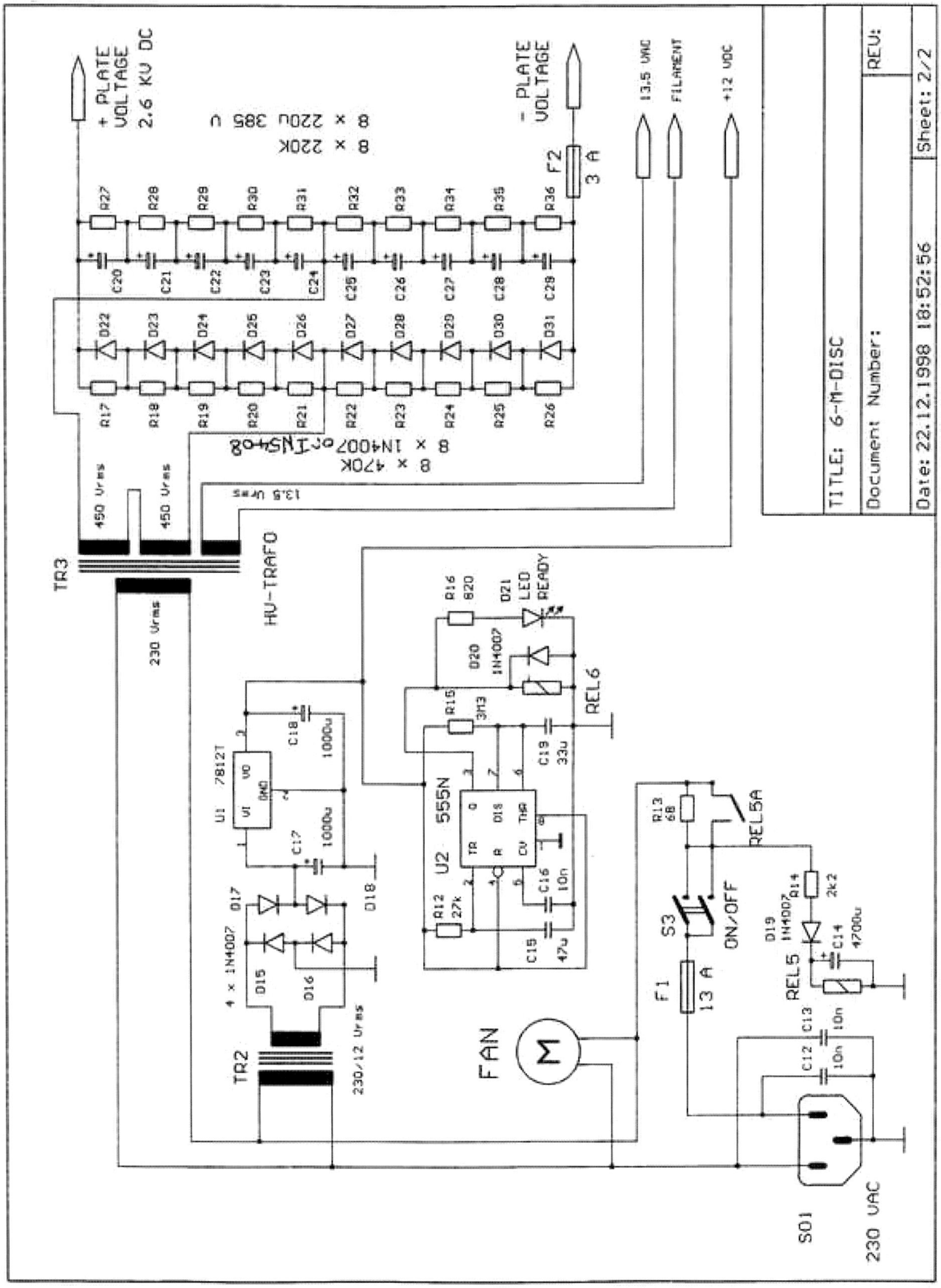

Bottom of the 6m amplifier

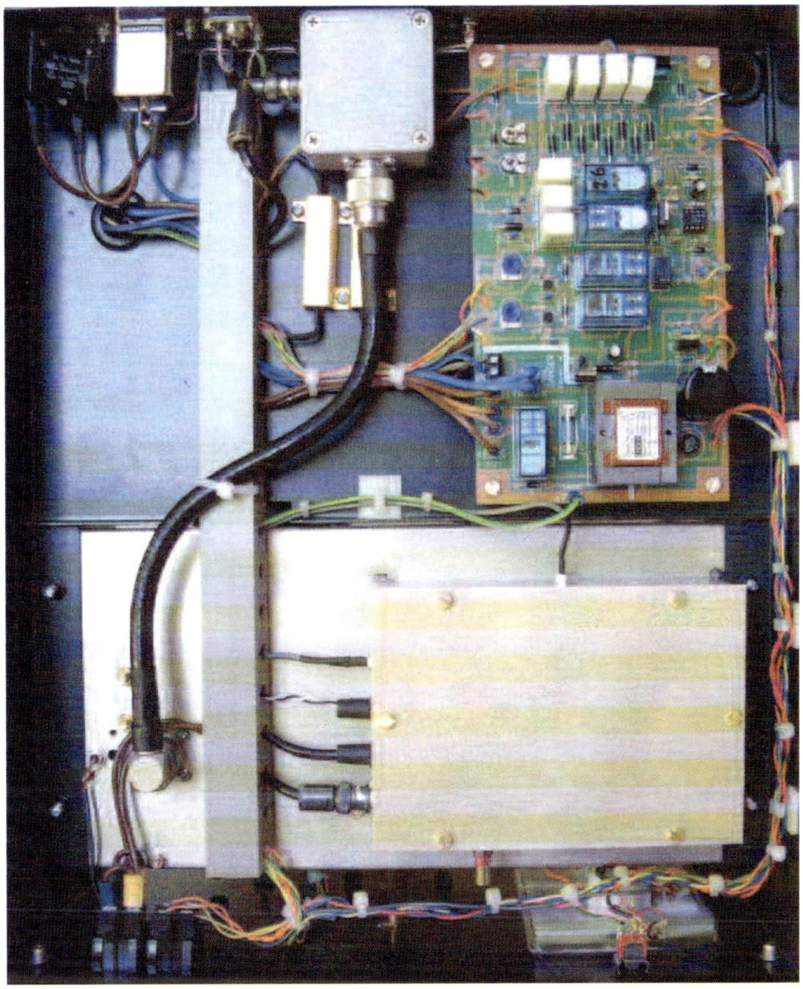

The above picture shows the component layout of the bottom of the 3CX800A7 amplifier. The input box is at the bottom right corner. At the top centre is the output relay which is a 24 volt single pole 20 amp Omron relay. It is mounted in a die cast box and will give good screening and handle the 1000 watts.

The G3ZRS/G3TEU control board is on the right. Mains input and circuit breaker are in the top left.

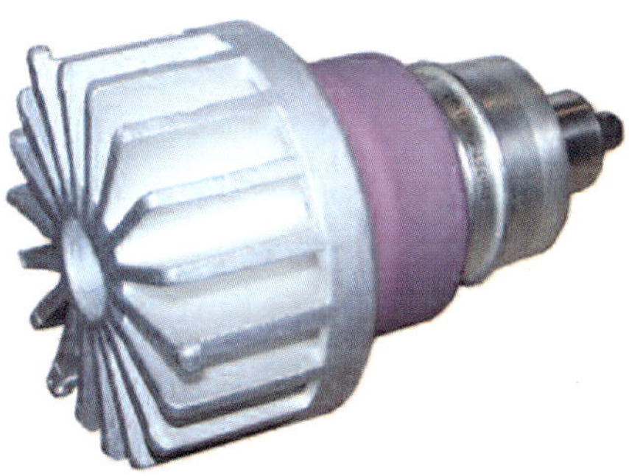

GS31b

GI46b

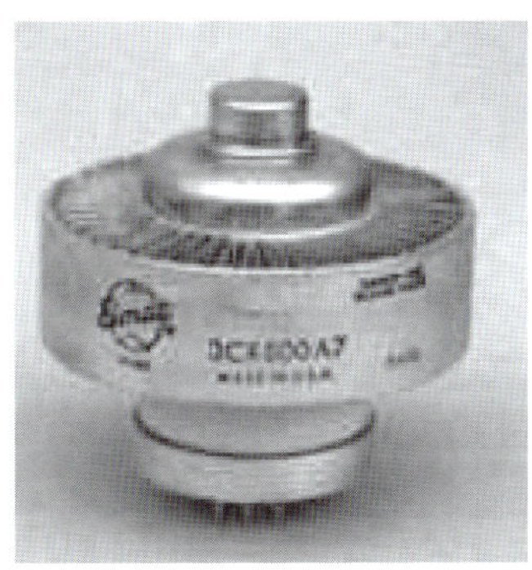

3CX800A7

Using the GS31b or GS 35b for 4m and 6m

Using either valve for 4m or 6m will make a very good stable amplifier. They are inexpensive and have a good specification. The GS31b is in many ways similar to the 3CX800A7 in performance. The GS35b is similar to the 3CX1500A7. The two valves do not have the gain of the 3CX valves. Now most radios deliver around 100 watts so you will not have a problem getting the drive power you require. Both valves use a 13 volt heater at 3 amps. This is only 39 watts but they do need a 3 min warm up time as they are an indirectly heated cathode. The body of both valves are very similar. The GS35b has a larger copper cooler and runs at around 3600 volts. The GS31b has a smaller metal cast cooler and runs at 2800 volts. The bias voltage for the GS35b is 36 volts and the GS31b is 28 volts. In the following descriptions both valve circuits will tune 4 and 6m. The T input circuit capacitor can be brought out to the front panel and marked 4m or 6m as the one tuned circuit will cover both bands. The PA coil can be switched by just shorting one turn out and that will give you operation on 70 MHz. Mark the front panel with the positions for each band.

The above picture shows the GS35b amplifier layout with the Pi L network clearly seen in the PA compartment.

The tune and load capacitors were made in one module from Cap Co., unfortunately not available any more. You will need a 50 pF variable for the tune and 200 pF for the load, not 50pF as shown on the circuit! Sorry. Due to the high voltage of 3600 volts the PSU has 10 x 220 µF 450VDC capacitors in the voltage power supply.

The Pi L network is 4 turns of 6 mm copper tube on a 30 mm former and the L section is 5 turns of 3mm copper tube on a 12 mm former.

The anode choke is 42 turns of 1 mm enamelled wire on a 20mm PTFE former.

The GS31b version of the 6m amplifier

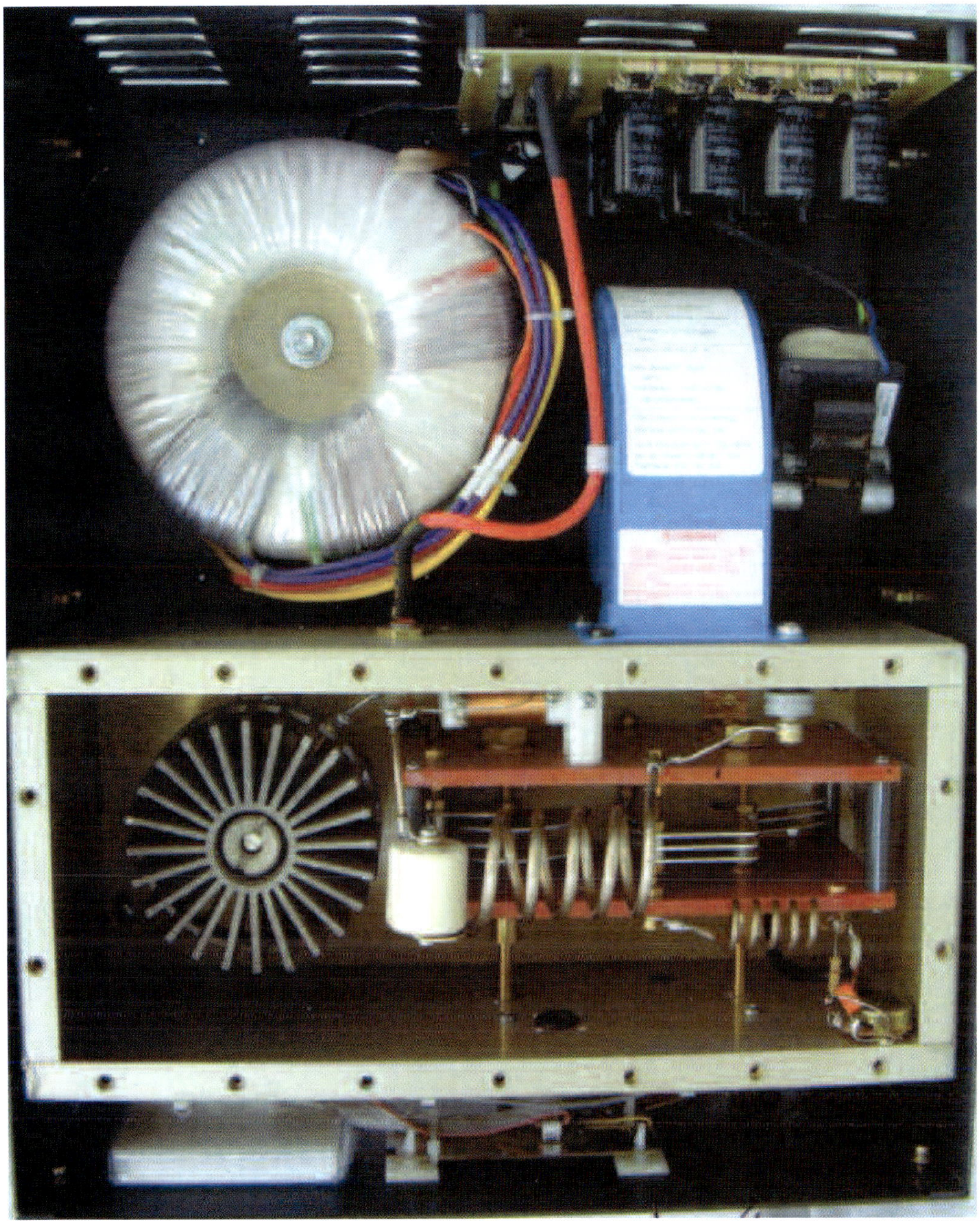

This picture above shows the GS31b mounted on the left of the PA box.
The Pi L network is mounted above the tune and load capacitor.
The DC feed is through a 680pF 8kV feedthrough capacitor.

The Pi L network is 4.5 turns of 6mm copper tube wound on a 30 mm former.
The L section is 5 turns of 3mm copper tube wound on a 12mm former.
The anode choke is 42 turns of 1mm enamelled wire on a 20mm PTFE former.
A load capacitor of 200 to 300 pF would be required

To use on 4m just use 1 turn less on the PA coil or short one turn with a ceramic switch to give dual band operation, sorry no picture available at going to press.

Input box of a GS31b or GS35b amplifier for 4 or 6m

The input capacitor will tune both 4m and 6m bands without modification. It is a normal T match input circuit.

The Heater/ Cathode choke shown at the top of the box is 2 x 20 turns on a 6mm ferrite rod. It could be replaced with 2 x 5 amp 40 μH ferrite chokes. The variable capacitor is 50pF.

The two input coils are both 10 turns of 1.5 mm enamelled copper wire on a 12 mm former.
The coax input relay is mounted in the bottom left of the box. It is a standard single pole Omron relay which is wrapped with silver foil and bonded to earth.

The GS31/35 base can be seen, you can solder directly to the centre connection which is the heater.
The next one up is the Heater/Cathode you will need to make a clamp as shown. It can be 6 to 10 mm wide and made of brass or steel banding.

The valve is fastened with four L shaped clamps made of 2mm thick brass and 25mm wide.

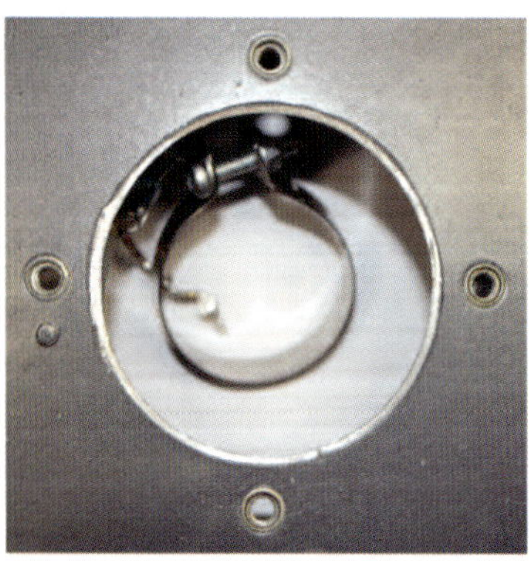

The above pictures show the fastening for the GS31/35.
4 of 25mm x 25mm x 7 mm L brackets clamp the valve into the valve hole

RF Stage GS31b/GS35b

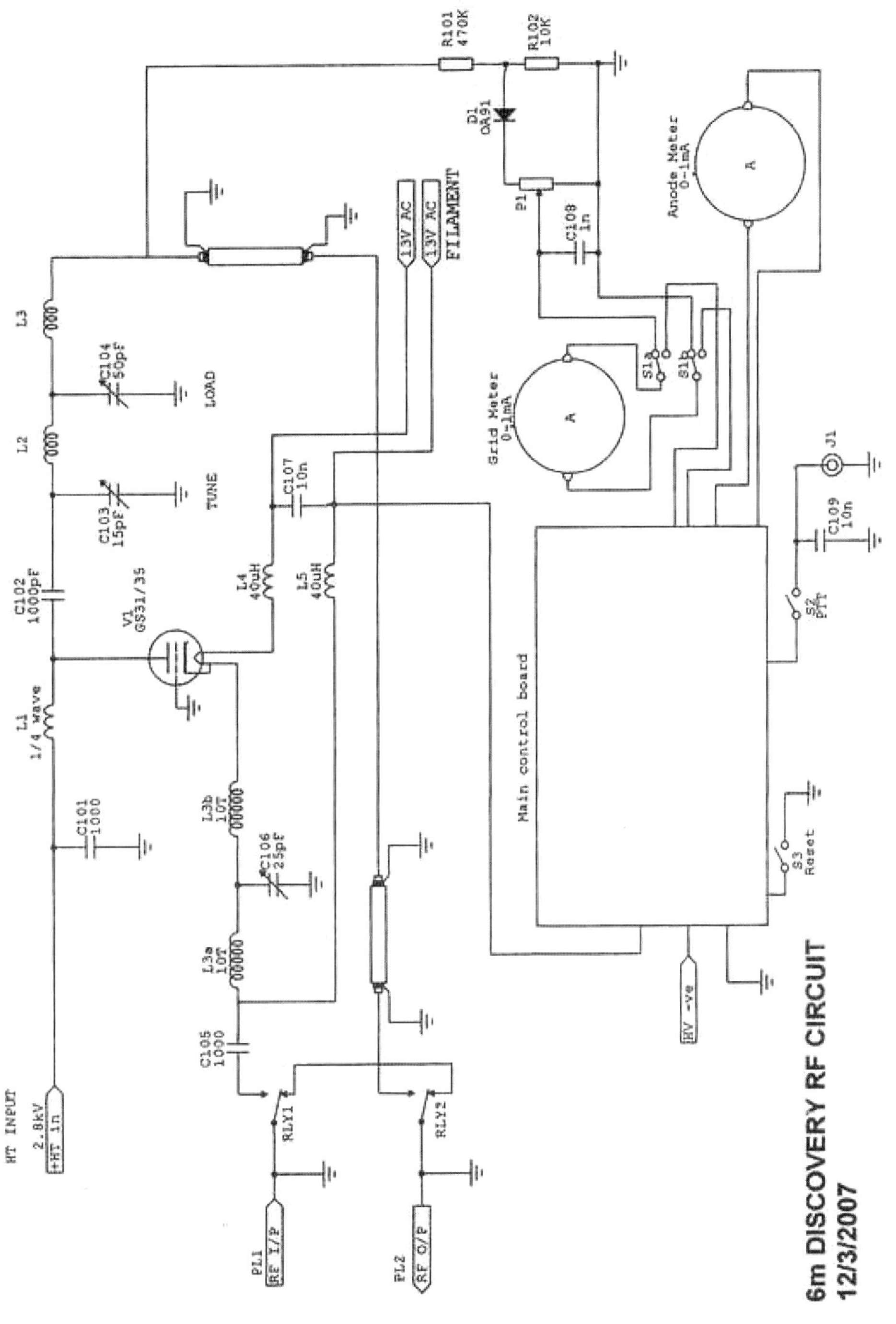

Note load capacitor should be 200pf not 50pf as marked unless you use an extra fixed capacitor.

Making a 70 cm amplifier and the choice of valves

Some 30 to 40 years ago there was not a lot of choice of valves capable of producing 300 to 500 watts at 70cm for the average radio amateur. A single 4CX250b would give around 250 watts and a pair would give over 400 watts.

Using the popular Russian ceramic Triodes

Russian GI7b, GI46, GS31b and the GS35b triodes came along at a very reasonable price.
They do not need expensive valve bases or special anode clamps and finger stock like other valves.
For this reason I only would recommend the Russian triodes, they will give you the power you require from 300 watts to 1500 watts, simple to make and stable in operation.

I have built several 1500 watt 70cm amplifiers using the GS35 but the metal work is much more complex than using the other valves like the GI7b, GI46, and GS31b. On the smaller valves the anode cooler can be removed and it leaves a smooth flat surface so you simply fasten a ½ wave brass line to it with a 10mm hole at one end of the line.
The GS31b amplifier I describe uses a cam to push the anode tuning vane in and out. This is so the RF box can go from left to right across the box. A simpler way is just to use a 5 or 6mm set screw with a block of PTFE between the set screw and tuning vane and push it in and out from the end of the box
You can fasten it between the valve anode and the anode cooler described later. Tuning is a single flapper that moves back and forth from the end of the tuned line. The output load is a variable capacitor coupling to an N socket on the side of the case.

The case is easy to make using two aluminium boxes with sealed corners either welded or inserting aluminium 15mm angle and screwed into the corners. One long one for the anode compartment and the other smaller one for the tuned input circuit (dimensions later in the drawings).
The anode tuning can be very simple by just using a set screw at one end of the box with a PTFE insulator as shown in the drawing. This is OK if the back of the RF compartment is facing your front panel but in my case this was not suitable as my cases were made to take the RF compartment from left to right rather than front to back. I developed a tuning cam, which I will describe later, which allows the RF box to go from left to right and have the tuning knob anywhere on the front panel.

The input circuit is a tuned line with a coupling loop running along side of it. It gives a flat easy to tune input of 50 ohms to your radio.
The grid ring for all 4 valves can be clamped directly into an aluminium chassis with small L shaped brass brackets. The tuned line is fastened directly to the anode and valve cooler which also helps the valve cooling at full power.

I must mention that there is a small amount of thermal drift when these triodes are used on 70 cm. This problem is not noticeable at all on 2m. I do not think it is anode expansion because using extra cooling or a thicker line does not seem to make any difference. I suspect it is the grid expanding closer to the anode when hot. The problem is not a major concern, when you tune up allow 20 seconds to pass at full carrier then tune for maximum output. The drop off is about 10% and as soon as you start talking the power comes up to full power.

A GS31b 800 watt 70 cm amplifier

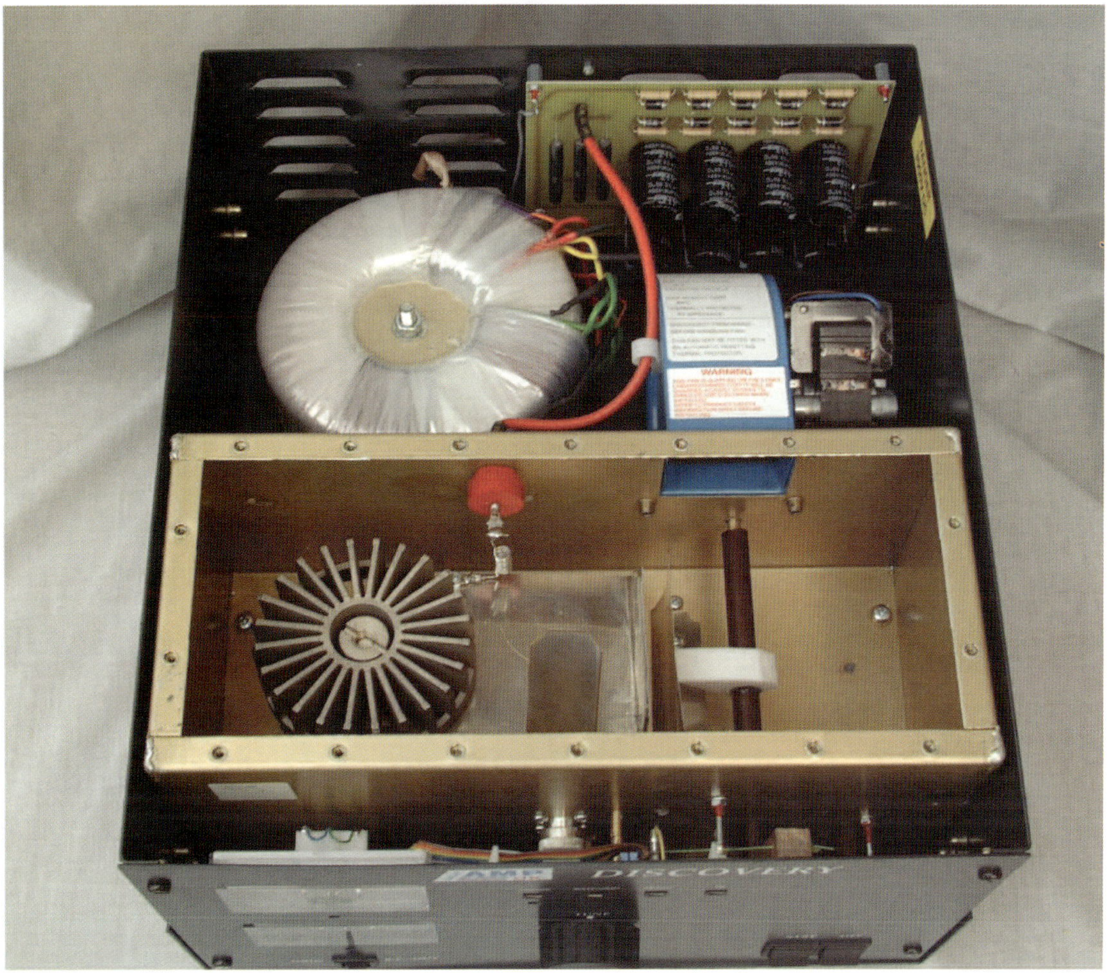

The picture above shows the tuning of the half wave line instead of using a set screw at one end. The brass tuning vane is fastened to the bottom of the box and is pushed in and out using a cam which is rotated from an arm at the front of the box which in turn goes to a slow motion drive on the front panel using 30kg fishing line.

The pushing arm is made of PTFE or Teflon and fastened to a 10 mm rod made of paxoline using a 6.5 mm brass rod inserted at each end, one goes under the fan as can be seen and the other extends by another 40 mm and another arm is fastened to this which is 100 mm long and pulled in and out using a slow motion drive fastened to the front panel.

The half wave line is simply clamped between the anode and the anode cooler and no other support is required if you use 1.5mm brass which also helps cool the anode as all the compartment is pressurised by the snail fan.

DC is applied to the tuned line via a 680pF 8kV feedthrough on the back side of the anode box and a quarter wave choke.

12 x 8mm holes are drilled into the tuned line so the air can pass through the tuned line into the cooling vanes on the valve cooler.

The circuit is stable and easy to build and set up. It is more of a mechanical challenge than an electrical challenge. The circuit is the same for the all the Russian valves except the anode line for the GS35b.

This picture shows the RF output and the arrangement of the slow motion drive and the arm which pulls the cam in and out to tune the anode tuning plate.

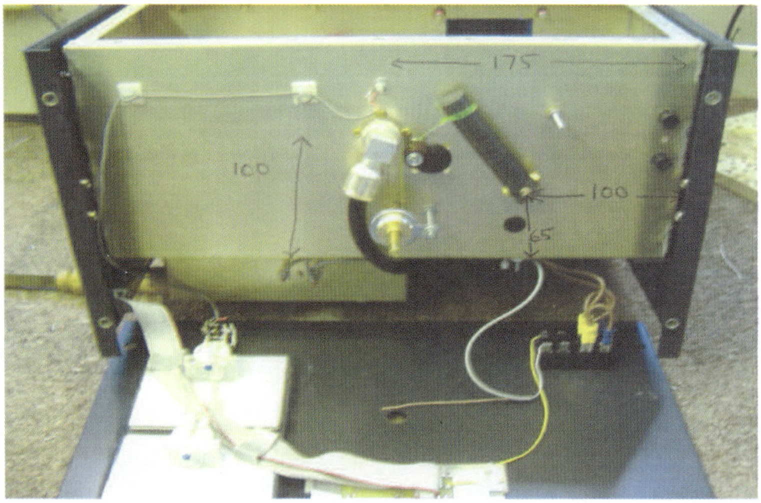

Inside view of the tuned input circuit of 70 cm amplifier which would be suitable for the GI7B, GI46b, GS31b and GS 35b

The tuned line is fastened as shown to the valve cathode and is resonated with the ceramic capacitor at the left end of the box.

The input link is fed through a BNC socket and adjusted with the coupling capacitor directly above the valve base.

The 13 volt AC heater supply is fed through the 2 chokes in the bottom of the box through 1000 pF 500V DC feedthrough capacitors.

The input coaxial relay need only be rated at 100 watts but the output relay most be well chosen to handle 1000 watts at 70 cm. There are Russian relays available at rallies like Friedrichshaven or on eBay and other Ukrainian and Russian web sites. You can also buy new from American and Japanese companies. Wimo in Germany and the DXshop in Wales keep a good selection of coaxial relays.

The 70 cm RF Circuit suitable for the GI7b, GI46b and the GS31b

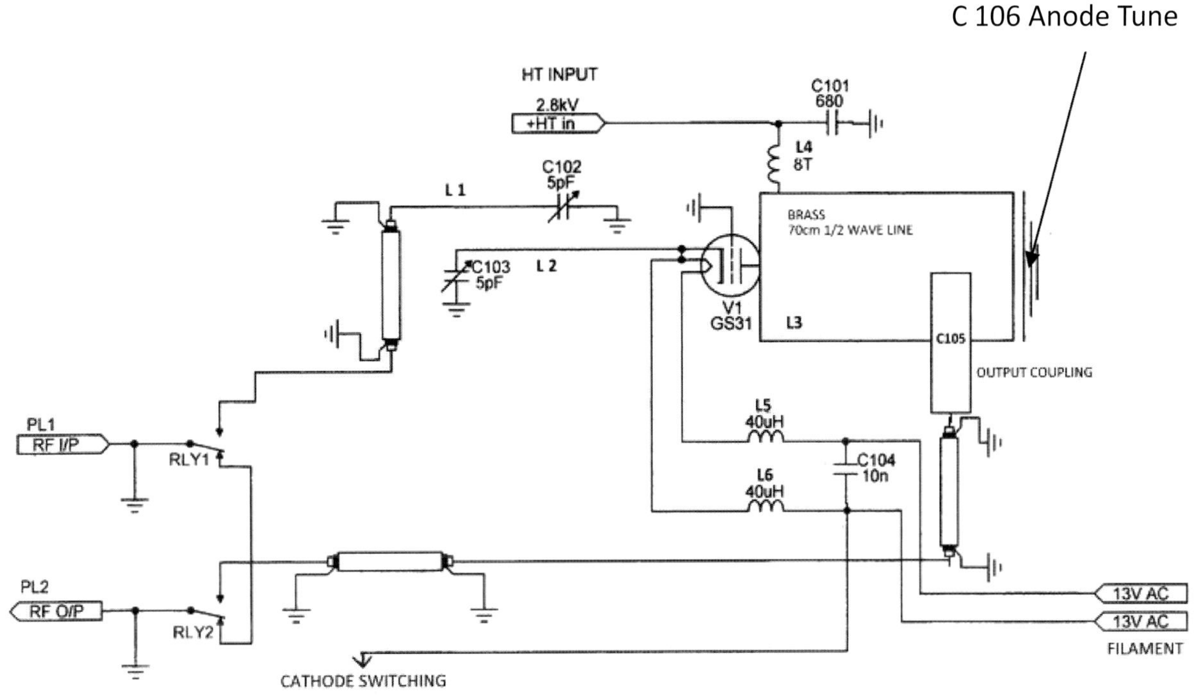

RF Components:-

V1 GS31b
L1 80 mm x 10 mm thin brass strip
L2 60 mm x 10 mm thin brass strip
L3 180 mm x 100 mm (see drawing) ½ wave line
L4 Anode RF choke of 8 turns on a 6mm former 0.5mm copper silver plated wire
L5, L6 Heater choke 8 turns on a 6mm former, 1.5 mm varnished copper wire
C 101 680pF 8KV feedthrough
C 102 5pF variable
C 103 5pF variable
C105 Output coupling, 80 x 30 mm, see drawing
C 106 Anode tune 100 x 65mm, see drawing
RLY1 1000 watt coaxial relay
RLY2 100 watt coaxial relay

N.B. GS31 is internally connected between heater and cathode

RF compartment of the 70cm GS 31b amplifier

Size of the Top RF box is 340 mm wide x 160 mm deep x 160 mm high
Size of the Bottom input box is 145 mm long x 100 mm wide 80 mm high

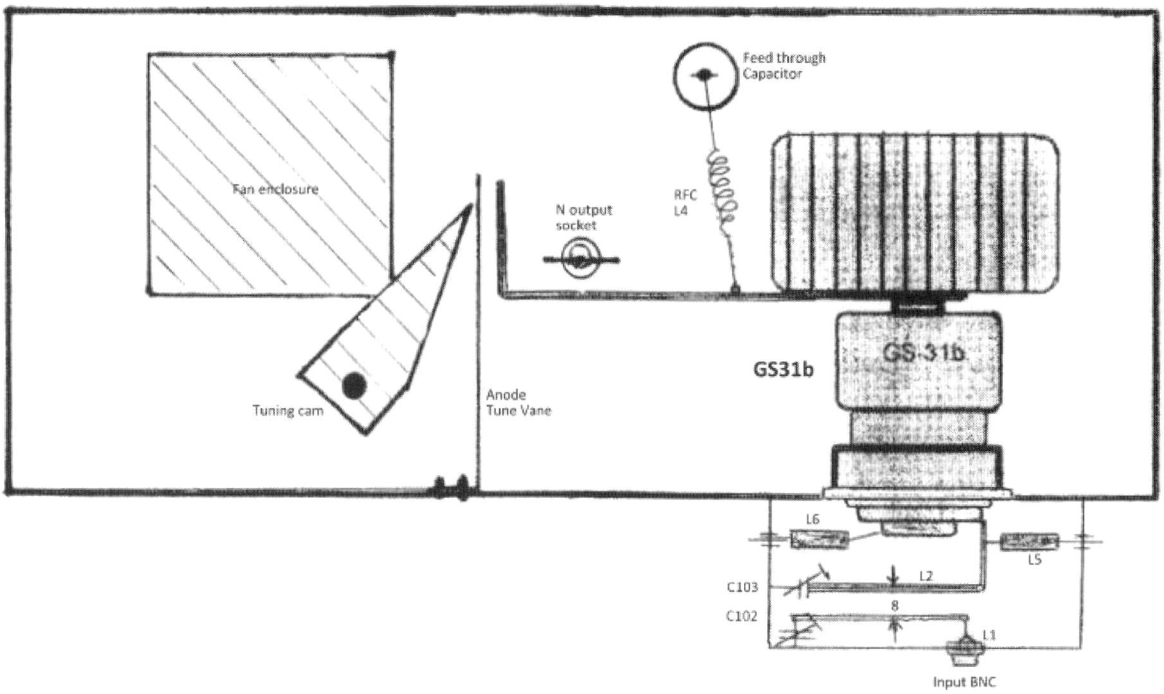

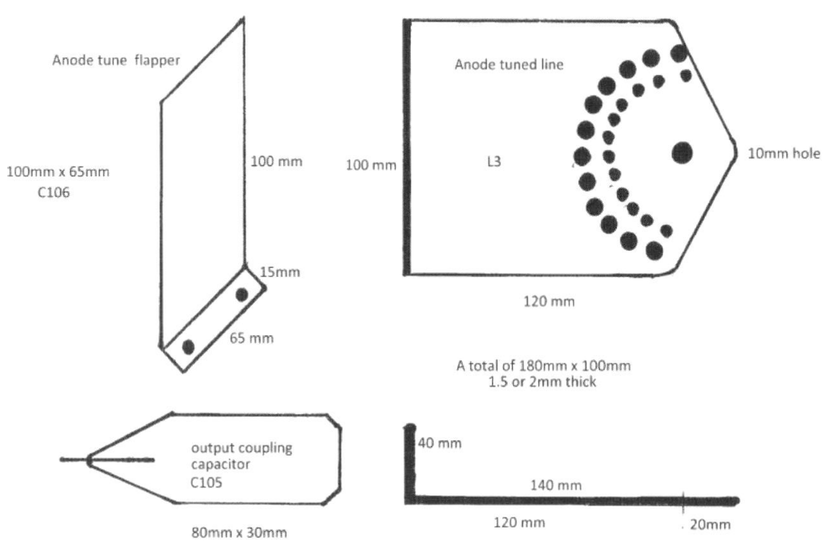

Output coupling capacitor C105

The output coupling capacitor is placed above the tuned line. You must fit a piece of Teflon or PTFE sheet to stop the capacitor touching while adjusting. The capacitor is made of 0.5 mm thickness brass and fastened to the N socket using a piece of 1.5mm wire soldered on the underside of the plate as shown.

Adjustment is from the top cover using a set screw and spacer. Alternatly use a pull cord made of 30kg fishing line and fix in the lid.

In practice

I find the adjustment of C105 does not make much difference. If C105 comes straight out horizontally from the N socket it is in the correct position and fine! I think putting the brass plate in the RF cavity is similar to putting a brass plate in the middle of a microwave oven. It gets hot wherever it is !

Size of the box

This is not critical. If you choose to use a simple set screw through the end of the box you can make the box shorter.

The top lid of the box as shown in the drawing below. An aluminium mesh must be placed between the chimney and the top of the box to stop any RF leakage. The chimney is a piece of 1 mm PTFE sheet rolled and fastened with small L bracket to the lid of the box to let the flow of air through the top of the box.

Alternative tuning method

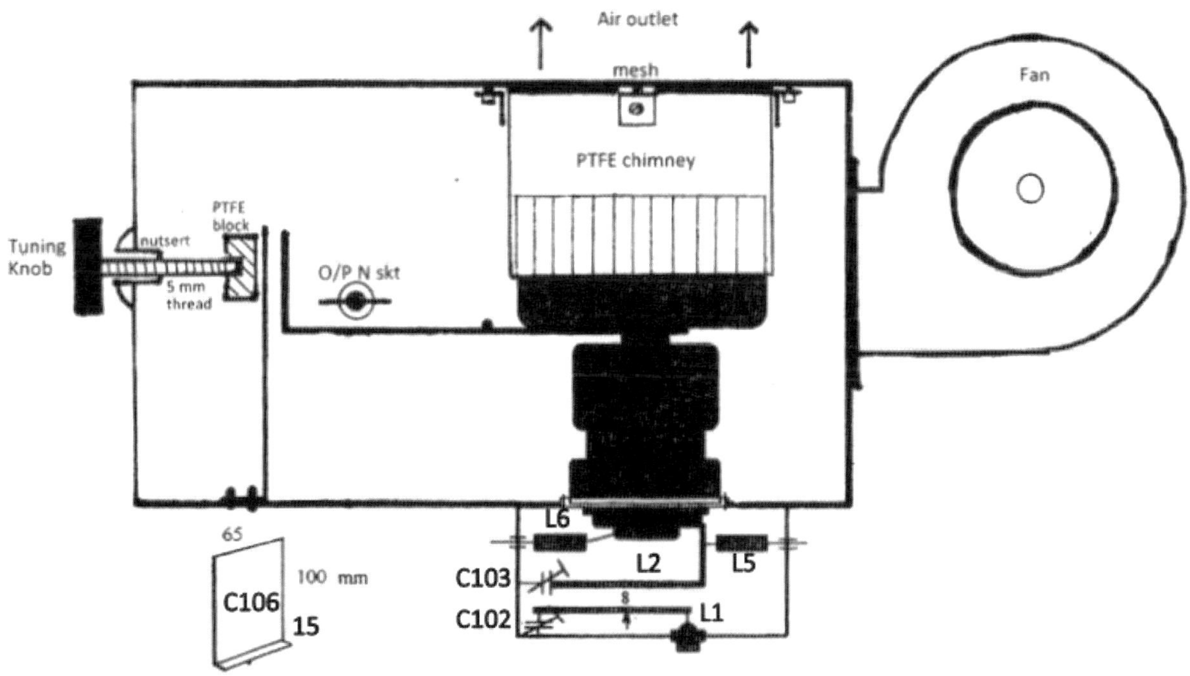

Above is the alternative method of making the RF box. It can be made shorter and the tuning is much simpler to make. The tuning knob end must be on the front panel, hence the box has to be running front to back in the main cabinet of the amplifier.

This drawing also shows how to mount the valve chimney. It is a sheet of PTFE rolled like a toilet roll and fastened together either with staples or 2 x 3mm nut and screws. It needs to cover half of the valve cooler and be a good fit. It is fastened to the top lid with small L brackets as shown.

Between the fan and box there should be a fine mesh which acts as an RF screen and is the true RF end of the box.

The tuning is a length of 6 mm threaded rod. It fastens into a PTFE block which should be drilled out to 8mm so it runs freely inside of the block. The thread goes through a 6mm nutsert which can be riveted in.

Alternatively you could tap a piece of brass and fasten the brass to the side of the box. You should also use PTFE between the tuning vane and tuned line, also between the tuned line and RF output capacitor. You can buy PTFE sheet which has a sticky back and is ideal to fasten on to brass.

Chapter 11
How to make HF amplifiers

Making an HF Amplifier

Before Starting

Before starting to make an HF amplifier you have to ask yourself a few questions:-
Should I use a Triode or Tetrode?
Do I want to make it easy or sophisticated?
What power would I like to achieve 500w 1000w or 1500 watts?
Do I already have a valve that I would like to use or should I think about buying one?
Shall I use a toroidal transformer or an iron core one?
Obviously most hams would like as much power, for as little cost as possible.

The choice of a valve to use

Traditional older valves

3-500Z, 4-400, 572b, 811, 813 are still around and readily available, these valves do not have a high gain but will make an easy to build amplifier for HF. You will not need a sophisticated control board and only one HT supply.

Newer Ceramic Triodes

The best Triodes are the 3CX800A7 and 3CX1500A7 both made by Eimac since the early 1970s. The cost of these are expensive for most ham radio home brew projects. The Chinese make both the 3CX800A7 and 3CX1500A7, the quality is good and at about half the price of the Eimac version. If money is no barrier I would use either a single 3CX1500A7 or a pair of 3CX800A7 which is what Alpha used throughout their years of production and is still used in the new RF Concepts 9500.

Russian Ceramic Triodes

Available from the internet or from large radio rallies like Friedrichshaven. They are easy to use and have a good gain compared with older glass valves. The GI7b or GI46b are a good start and a very good choice for a medium power amplifier, by using 1 or 2 in parallel. They are similar to the 3CX400A7/8874 and can be a good substitute.

The GS31b is similar in specification to the 3CX800A7, a bit larger and not quite as much gain, but excellent value for money. Like the 3CX800A7 it will give 1 kW.

The GS35b is the largest of the Russian triodes. It has a similar specification to the 3CX1500A7, a slightly lower gain but excellent value for money for £100 to £ 150. A 100 watt radio will easily drive the GS35b to over 1500 watts output.

None of these are made any more but are still available in good quantities and often called NOS, new old stock.

Modern Ceramic Tetrodes

4CX800/GU74

The most common tetrode over the last 10 to 15 years has been the Russian GU74 or the western number is 4CX800. These are also NOS, new old stock and have not been made for many years. Very strange that new amplifiers are still being made using these valves that are obsolete and no longer made anymore. They are cheap compared with other western alternatives.

4CX1000

The 4CX1000 is the nearest available equivalent. The 4CX1000 became available around 1948 as a popular valve for commercial service and also used in amateur service in the famous Collins 30S1.

The new RF Concepts 8410 uses the 4CX1000 instead of the GU74 which was used earlier in the Alpha 91b. The 4CX1000 uses a special bayonet valve base, these are sometimes hard to come by, but otherwise will make an extremely good amplifier.

GS23B/4CX1600U

The GS23B is the latest of the Russian tetrodes to become available and is now being used in the latest OM 1500 watt amplifier.

It does look an excellent tetrode and the cost is also very reasonable. It is stocked by European and American suppliers. I have just looked on eBay and amongst many others you could buy one for 99 US dollars. Like other Russian valves I am not sure how old they are. Maybe like other Russian valves NOS, new old stock. In many ways it looks to have a specification similar to the 4CX1000.

Triode or Tetrode

Triode - Glass

There are many pros and cons with either choice. The first advantage with a glass triode is it only requires one HT supply, and a simple control board. Usually these are the cheapest type of valve to buy. The gain is around 10dB. A pair of 3-500Z will give you 1000 watts with 100 watts of drive.

Disadvantages

40 to 50 ago years most amplifiers only worked on 5 bands, 10 to 80 meters. Now most amplifiers can work on 11 bands 10 to 160m, WARC, plus 60m and 6m. A grounded grid amplifier, glass or ceramic needs a tuned input circuit to match the incoming coax to the valves cathode circuit. This means every band has to have its own tuned input circuit which has to be wired, switched, and tuned.

Triodes - Ceramic

Despite the price the 3CX800A7 or 3CX1500A7 are excellent valves with a gain of up to x 40. You will only need 10 watts of drive to get 400 watts and 25 watts will give you 1000 watts. This is fantastic for a small triode and stable in every way.

Russian ceramic GS31/35 triodes are one of the best choices you could make when starting to think about making an amplifier. The price is a fraction of the 3CX series and you do not have to worry about a valve base. The valve grid can be clamped straight into the amplifier chassis. Their gain is not as high as the 3CX series but most modern radios have 100 watts of power available so the drive is not a problem. The GS31/35 will work well on HF and 50 MHz as the upper frequency is above 500 MHz

Disadvantages

The cooling of the valve has to be ducted through an appropriate chimney, so that the air passes through its anode cooler. They require more metalwork to create a pressurised compartment, also a higher pressure cooling fan.

The 3CX800A7 and 3CX1500A7 are expensive and pullouts are hard to come by. Chinese equivalents are available at a better price. The 3CX1500A7 has a cheap ceramic base. The 3CX800A7 has a more expensive Johnson 11 pin special base.

GS31b/35b

The price is good and with no complications. Availability is good but all NOS, new old stock, and not made anymore but reserves of the valves are good. The bias voltage is higher than most triodes at 28 volts and 36 volts but this can be easily done with a transistor regulator or zener diode.

Tetrode - Glass

Glass tetrodes are not around any more!

Most circuits that have used the 4-250 or 4-400 have strapped the grids and made them into the equivalent of a triode. The gain is low when this is done and the valves are large in size.

Tetrodes - Ceramic

These are a good option and I would seriously consider them. They have definitely been rejuvenated as in the 1950 and 60s the 4CX250b was so popular for everything VHF and HF, and the larger 4CX1000 from 1957 really transformed all larger amplifiers amateur and commercial.

The GU74/ 4CX800, GS23b/4CX1600b and 4CX1000 all will make a very good 1 kW+ amplifier for HF. You will still have all the usual concerns with a tetrode. A regulated screen supply, a variable grid bias, start up timer, grid and anode over-current protection, sequential switching but all this can be simply done with a GM3SEK Tetrode control board at a very modest price, do look at his web site.

Gain

Modern ceramic tetrodes for their size produce a large power with a low level of drive. With an external anode cooler they run very cool. They have a robust filament that hardly ever goes open circuit compared with glass valves. Their gain is high but hardly ever used to their full potential. All modern tetrode amplifiers feed a 50 ohm load resistor. Traditionally a tuned grid input circuit would drive the valve to its full power potential with only a few watts of drive.

Russian valve bases are available at a fair price. The 4CX1000 needs a more expensive bayonet type. Tetrode valve bases have decoupling rings built in to improve stability.

All modern tetrodes use an external anode cooler so an appropriate anode chimney has to be used so the cooling air passes through the anode cooler.

Protection

Safety protection must always be fitted. The bare minimum is the transformer must be fused on its primary. An amplifier that gives 1000 watts will have an approximate input of 2000 watt which is 8 amps off a 240 volt supply, or 18 amps off 110 volts. An easy figure to remember on 240 volt mains, every 1000 watts taken is 4amps.

Due to all the high voltage components in an amplifier, any one can break down and cause a large power surge. More sophisticated amplifiers trip the HT transformer out until the fault clears.

HT fusing is difficult due to the high voltage involved, if the fuse goes it will try and track over a normal fuse.

An HRC fuse would be better where the sand inside melts around the wire.

A better idea is to fuse the negative of the power supply.

Adding current limiting is a good idea in the HT supply, 15 ohms per 1000 volts will help stop damage to the transformer or power supply in the HT supply. These must be of a high wattage of around 15 watts.

Planning the inside of the cabinet

It is very well worth giving the component layout plenty of thought and time.

Before you start drilling holes in your new case, I have always got all the major components and put them onto a large piece of cardboard card the same size as your proposed cabinet. Try moving them around until you think everything looks well, mark the card and use it as your template.

Isolation and switching

Isolation is an important part of building an amplifier to ensure you get stability and reduce the chance of taking off. Generally HF amplifiers use a double pole 20 amp relay for the coaxial in and out switching. If you want quicker switching and then you will have to purchase two single pole vacuum relays.

The main concern is to keep the input circuits away from the output circuits whether you are making a tetrode or triode amplifier. The input circuit is often below the chassis and the output above. Alternatively the input can be built into a separate box as long as you can get to it for adjustment purposes.

Power Supply

The power supply is better kept away from any high RF field and convection heat from the valve. Only two items in an amplifier produce any amount of heat that is the valve and the transformer. It is worth passing the air from the fan across the transformer first before it passes through the valve.

The transformer is the heaviest part of any amplifier and many large amplifiers ship the transformer separately, so the case does not get damaged in transit. If you use steel for your case it should be fine to support your transformer. If you use aluminium, especially if it is only 1 or 1.5mm thick, it must be strengthened by another piece of plate. Both toroidal and iron core transformers can distort a case when being moved around. When you have drilled the mounting holes for your transformer, it is worth leaving the transformer out until you have almost finished it. It can be quite a hulk to move around while you are still building.

Transformer

It is your choice but I would now always choose a toroidal transformer, they are lighter, smaller and usually cheaper than the iron core types. Most companies will wind you one to your specification. In the UK try Roger Banks at the DXshop in Wales. He keeps a good range of transformers and other parts you will need for your project.

PSU

The power supply board is better kept together as one unit. The diodes, electrolytic capacitors, balance resistors, are easily mounted on one PCB made by yourself. Companies in the USA as well as the DXshop in Wales sell complete power supply boards.

Cooling

Glass valves

Glass valves like the 3-500Z and 572b do not need a high pressure cooling fan. Air is more quantity than pressure. Axial fans like Papst 120 mm and others make a good choice for cooling. The motor size needs to be between 10 and 20 watts. When you choose it, first check how noisy it is, some can be very noisy.

When cooling the 3-500Z and similar valves it is important that a certain amount of the cooling air passes through the valve base. Valve pins can become de-soldered!

Ceramic valves

Ceramic valves all employ an external anode cooler like a radiator. This is far more efficient than the convection cooling used with glass valves.

When choosing a fan for a ceramic valve it is important to look at its air requirements. Most specifications are for the valve running at 100% carrier. For SSB and CW the fan can be down rated by 50%.

Most ceramic valves do not need much air passing into the lower part where the base is. It is however a good idea to keep a small amount of air passing through the input area. This can be done with 2 x 10mm holes from the pressurised anode compartment into the bottom input box making sure that there is a way out through the input box for the air.

Making a simple 800 watt Triode Amplifier

Using

4 x 811 or 4 x 572b

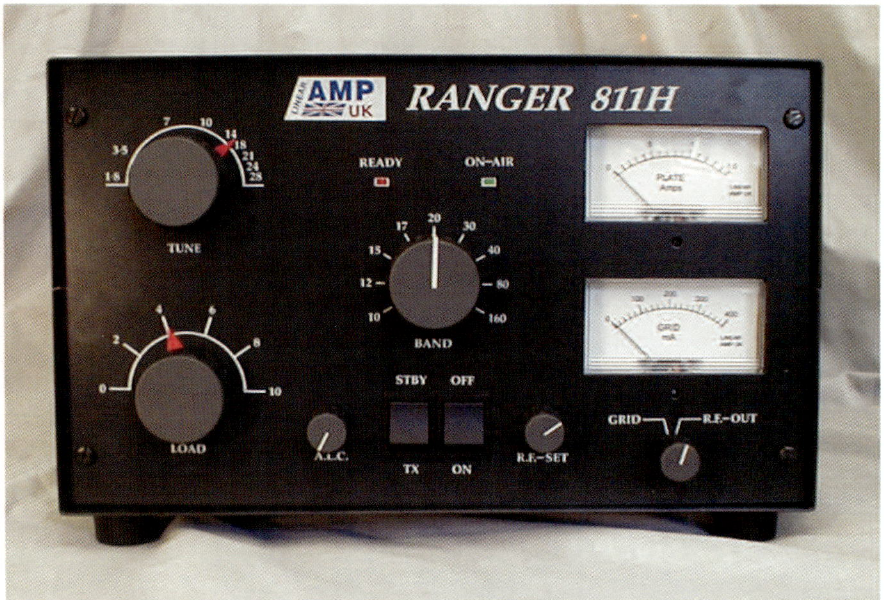

Front Panel layout of the 811 amplifier

Case size	335mm	wide	240mm	high	404mm	deep
	14 in	wide	9.5in	high	16in	deep

One of the easiest amplifiers to build is to use 4 x 811 triode or 572b triodes. Both valves are basically the same but the 572b has a carbon anode rather than the pressed metal one used in the 811.

I designed the Ranger 811 as a reasonably priced 800 watt amplifier after seeing a Collins 30L1. I sold the Ranger 811H readymade and also in a kit form called the Ranger 811K.

The 811 runs at 1800 VDC and if you choose the 572b you will not get any more power. The anode dissipation is higher so the valves run cooler. You can increase the HT to 2400 volts when using the 572b and you will then get 1300 watts! But I would not recommend it. You will have to fit neutralising which takes a small amount of RF from the anodes and feeds it back to the cathode circuit. Not difficult but it does need setting up and adds more complication.

I include suggestions of how to lay out the cabinet and major components. Most parts are available including the cabinet and transformer from Roger Banks at the DXshop in Wales, do look at his website.

The 4 x 811 in parallel gives a comfortable 800 watts for a relatively low anode voltage of 1800 VDC. All the tuning is quite simple. The anode load impedance is quite low using the 4 valves in parallel.

Bottom view of the 811/572b Amplifier

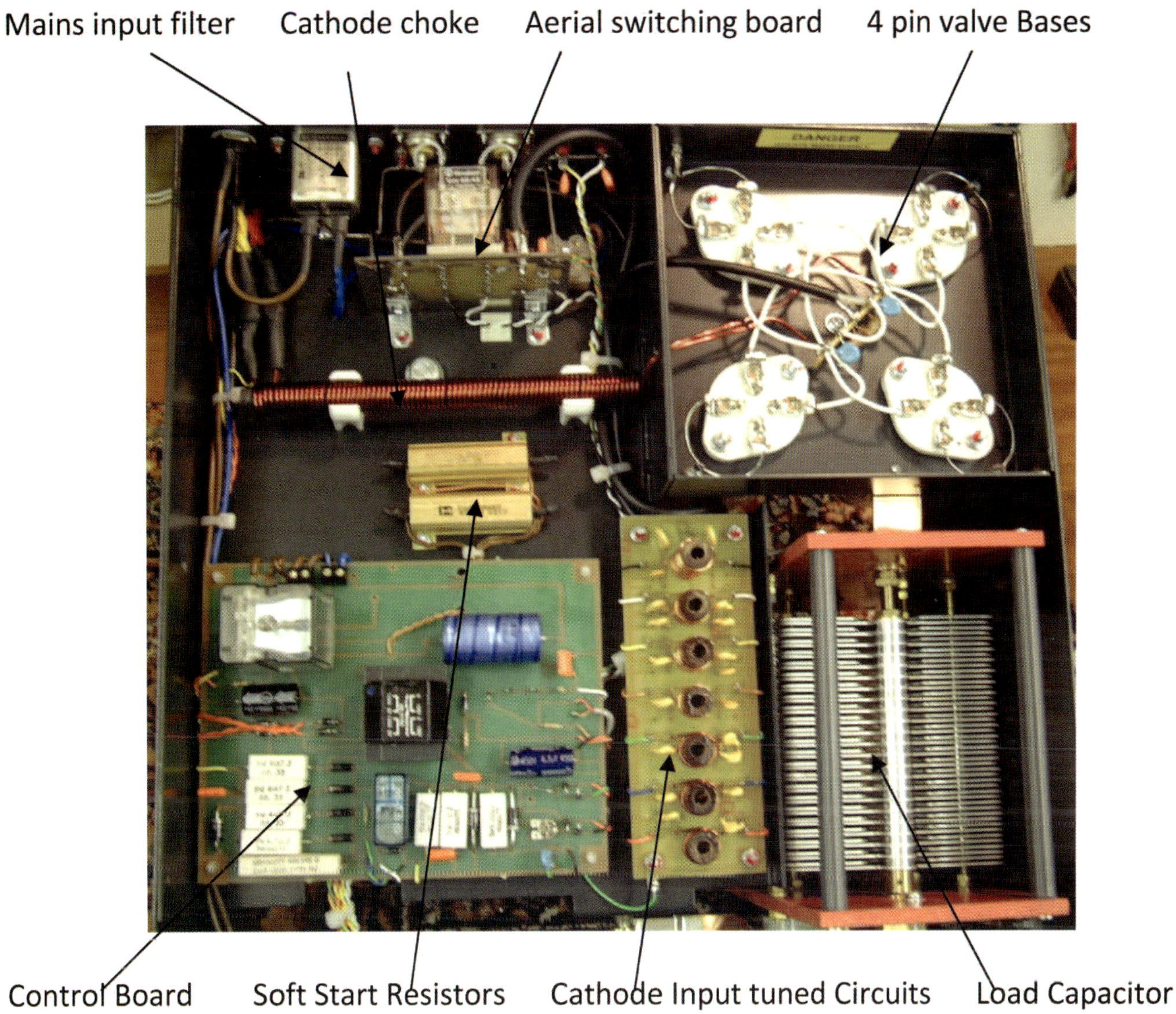

Control Board

811s and 572b do not need a sophisticated control board, this one has soft start, relay control, output for the indicator LED and ALC plus a 12 volt PSU for the relays and LED. See the circuit diagram details, easy to hard wire.

CONTROL BOARD WIRING (anti clockwise from top right)

1. RED+ve/YELLOW - to READY LED
2. RED/RED - to 12V on toroidal transformer
3. RED/RED - to Plate meter lamps (save 10cm trimmings to link to Grid meter lamps)
4. YELLOW - to Filament choke
5. BLUE+ve/GREEN - to ON-AIR LED
6. BLACK/WHITE- to Antenna Relay board
7. YELLOW - twisted with yellow wire from PTT socket and then both wires go to STBY/TX switch on front panel.
8. GREEN - connects to earth terminal on Input board
9. BLACK - connects to negative terminal on PSU board
10. RED+ve/BLACK - connect to Plate meter (save 10cm trimmings to connect Grid meter to Grid/RF switch)
11. RED/GREEN - connect to RF/GRID switch
12. BROWN/RED/BLACK - to ALC potentiometer
13. WHITE - twisted with white wire from ALC socket and then both wires go to ALC switch on front panel.
14. BROWN/BROWN - to 2K2 resistor
15. BLUE - to transformer
16. BLUE - to fan
17. BROWN - to fan
18. BROWN - to transformer
19. BROWN (med) - to 47R resistor
20. BROWN (large) - to ON/OFF rocker switch

1. Control pcb (fitted with 2 x solder pins and wire link)
2. 1 x 24V relay
3. 1 x 12V small relay
4. 1 x transformer
5. 1 x 4700uF electrolytic capacitor
6. 1 x 1000uF " "
7. 1 x 10uF " "
8. 3 x 1N5408 diode
9. 4 x 1N5338B zener diode
10. 11 x 1N4007 diode
11. 3 x 1N4148 diode
12. 4 x 4R7 ceramic wirewound resistor
13. 1 x 82R " " "
14. 1 x 22K " " "
15. 1 x 1R " " "
16. 1 x 0R33 " " "
17. 2 x 820R carbon film resistor
18. 1 x 10K " " "
19. 1 x 27K " " "
20. 1 x 3K9 " " "
21. 1 x 500R potentiometer
22. 4 x 10nF (.01uF) polyester capacitor (orange)
23. 1 x 10nF (.01uF) ceramic disc cap (blue)
24. 2 x 2 way terminals
25. 1 x 60cm thick brown wire
26. 2 x 30cm medium brown wire
27. 3 x 60cm thin brown wire
28. 8 x 60cm red wire
29. 3 x 60cm yellow wire
30. 2 x 60cm green wire
31. 1 x 15cm green wire (earth connection to input board)
32. 1 x 60cm blue wire
33. 2 x 60cm white wire
34. 4 x 60cm black wire
35. 4 x pcb mounts

MAINS WIRING

36. 2K2 50W aluminium wirewound resistor
37. 47R 50W " " "
38. 4 x M3 c/s bolts, 4 x M3 nuts
39. 1 x 60cm blue wire 1.5mm
40. 1 x 30cm blue wire 1.5mm
41. 1 x 60cm brown wire 1.5mm
42. 1 x 15cm brown wire 1.5mm
43. 20cm wide heat shrink
44. 15cm narrow heat shrink
45. 2 x ring terminals
46. 2 x yellow spade terminals
47. 2 x blue spade terminals
48. 6 cable ties

Power Supply - 811

The valve heaters are wired in parallel the same as connecting 4 light bulbs together. The heaters are 6.3 volts at 4 amps and fed from the cathode choke. This stops RF getting back to the heater transformer. In this amplifier the heater winding is on the same transformer as the HT winding. The Transformer is a 900 VA toroidal transformer which is very adequate for the job and never gets warm even at full power.

Power Supply Board

The power supply is fed from the toroidal transformer. The transformer supplies 650 VAC into a voltage doubler board which brings the HT voltage up to 1800 VDC. 6 x 400 volt 220 microfarad electrolytic capacitors are used for smoothing. The full circuit is shown later. The power supply board conveniently mounts directly above the toroidal transformer.

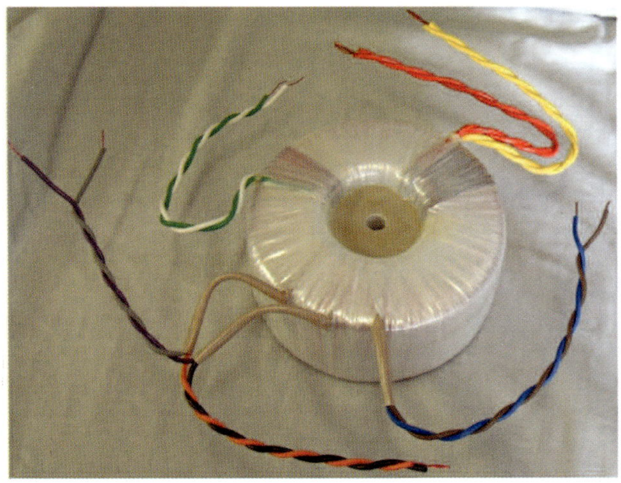

Above shows the Power supply and matching transformer. Available from the DXshop in Wales

The transformer windings are: -	Primary	0 - 230 volts Blue / Brown
	Secondary	650 volts Purple and Grey /Orange and Black
Centre tap	Heater supply	6.3 volts Red / Yellow x 2
	Relays and LED	12 volts Green/ White

Top view of the top of the 811/572b amplifier

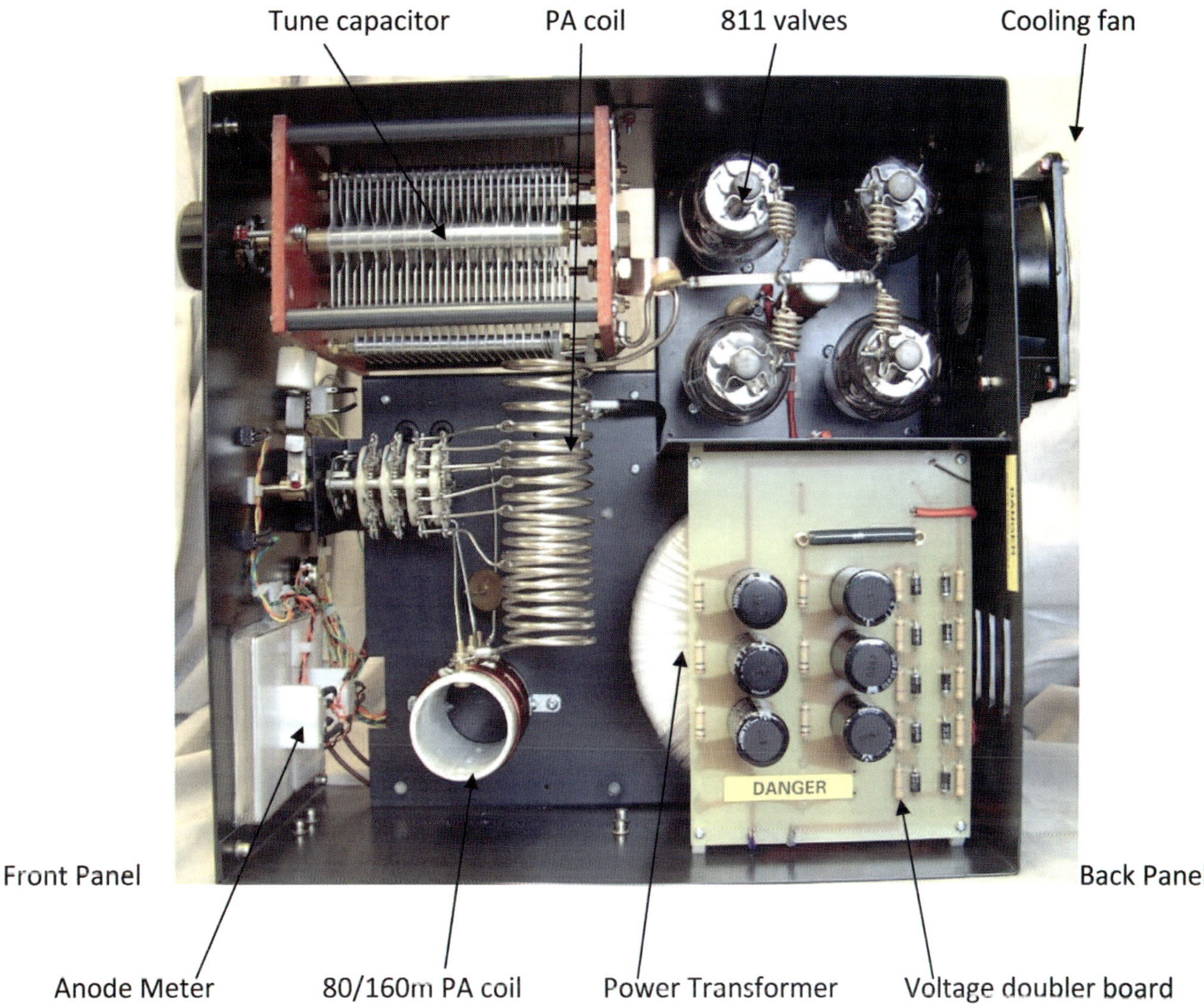

Top View of the 4 x 811/572b valves

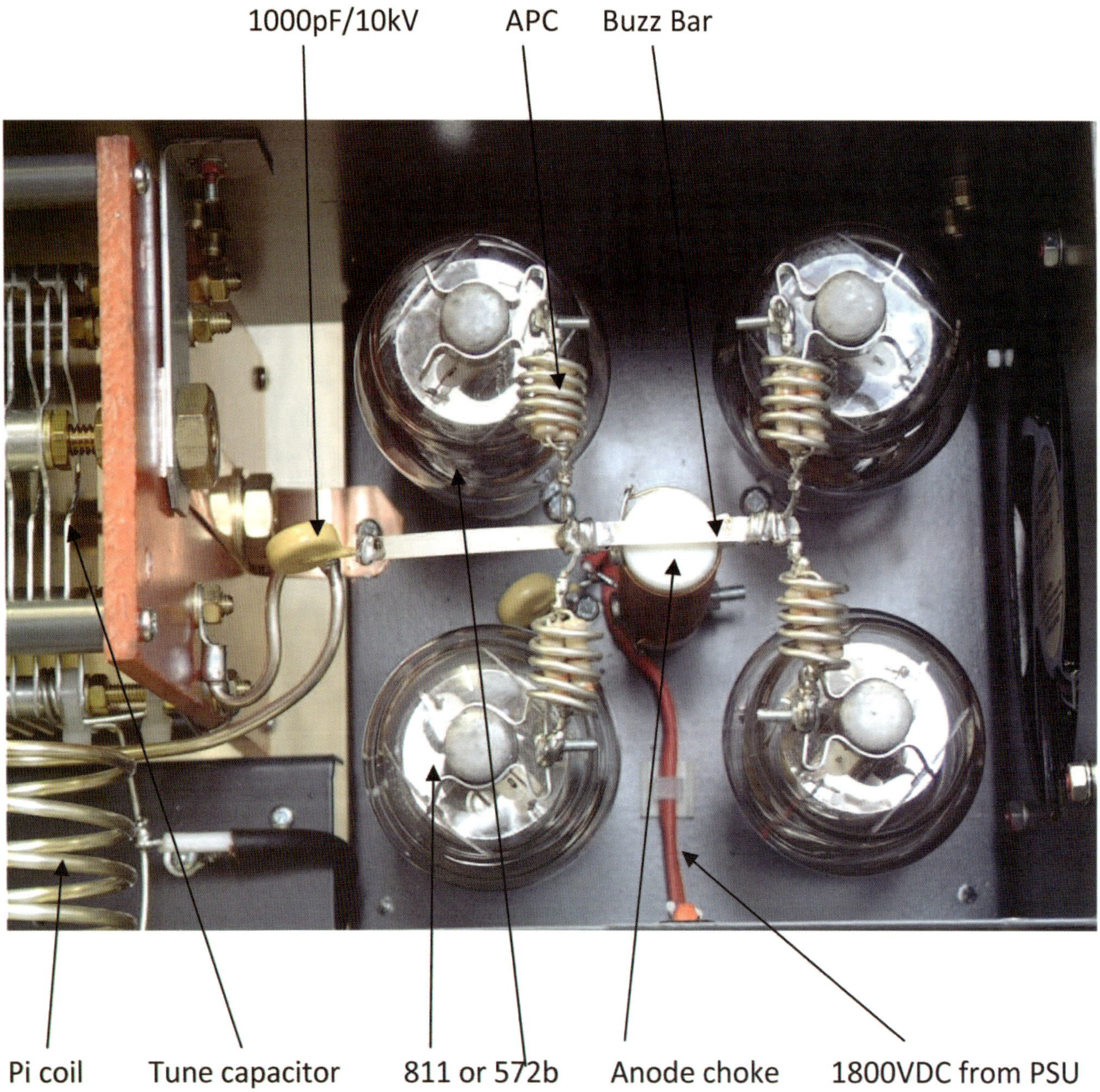

This close up shot shows the top of the valves and the anti parasitic chokes. The clips used are terry clips available from DIY shops often used for mounting tools etc on a wall.

The APCs are made of 1.5 mm silvered wire and wound on a 10 mm former. 3 x 150 ohm parallel resistors are fastened on the inside of them. The yellow blocking capacitors are 1000 pF /10kV. A loop is made in the PA coil and the capacitor mounted between that and the buzz bar that runs through the middle of the valves. The Buzz bar is mounted on the top of the anode choke.

Valve bases

The picture below shows a close up view of the heater wiring which is in white.
The grids and spare pins are earthed as shown.

Drive to the valves is fed through 2 x 10n capacitors.

4 pin valve bases

Front View of Ceramic Switch

Tune and Load Capacitor   Extra 300pF for 160m   PA coil 10 – 40m   PA coil 80/160m

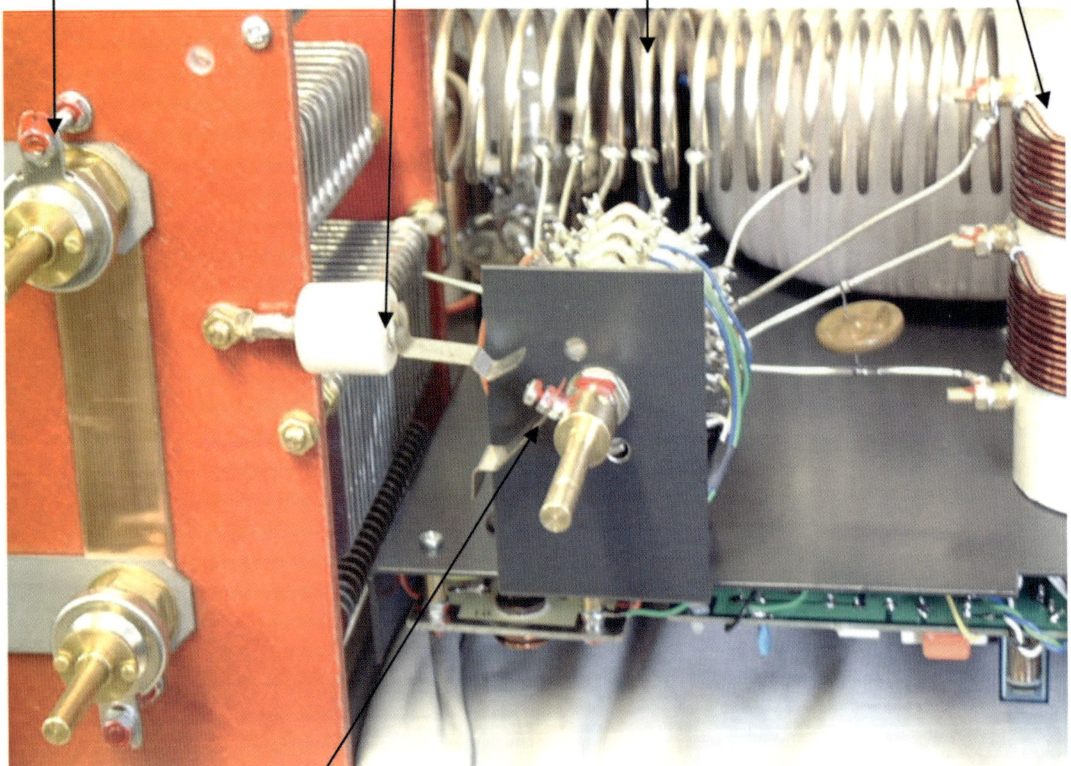

My method of switching the extra 300pF fixed capacitor for 160m.

Normally another wafer would be used. This method avoids that and the 300pF capacitor is permanently mounted on the main tune capacitor and earthed by the leaf switch mounted on the shaft of the ceramic switch when the switch reaches 160m. This method is used on all of my HF amplifiers.

PA coil
Coil winding Details

10 to 40m coil
3mm copper tube silvered if possible but not compulsory
1.5 inch / 38mm former
17 turns total tapped at

0	3	4	5	6	8	12	17	connected to top of 80/160m coil
Tune cap	10m	12m	15m	17m	20m	30m	40m	

80 and 160m coil

Former: - Fibre glass 110 mm long x 50mm wide
Wire: - 1.5 mm varnished copper wire
23 turns total tapped at

0	11	23
40m	80m	160m

Main Ceramic Switch

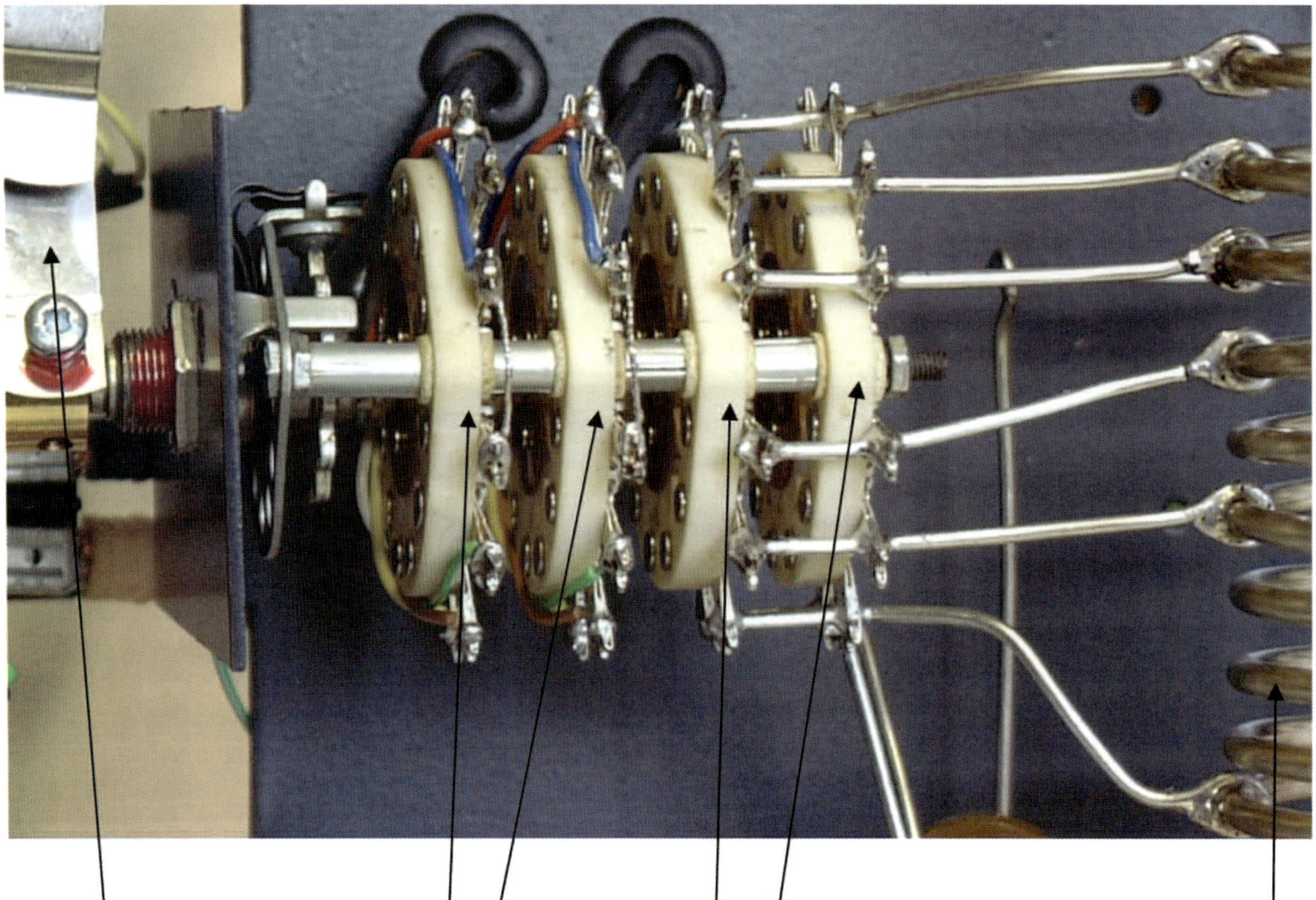

Tuned Input circuit board

The coils are wound on 10mm paxoline rod. A small ferrite slug slides in and out of each coil to resonate each coil onto the middle of each band for lowest input SWR. Each section also acts as a band pass filter.

Input tuned circuit

	10/12m	15/17m	20m	30m	40m	80m	160m
Input Capacitor	100pF	270p	330p	330pF	470pF	570pF	1000pF
No. of turns Wire thickness	4 (1mm)	5 (1mm)	7 (1mm)	8 (0.5mm)	10 (0.5mm)	17 (0.5mm)	27 (0.5mm)
Output Capacitor	100pF	150pF	220pF	220pF	470pF	470pF	1300pF

Aerial relay board, cathode choke and the back panel of the amplifier

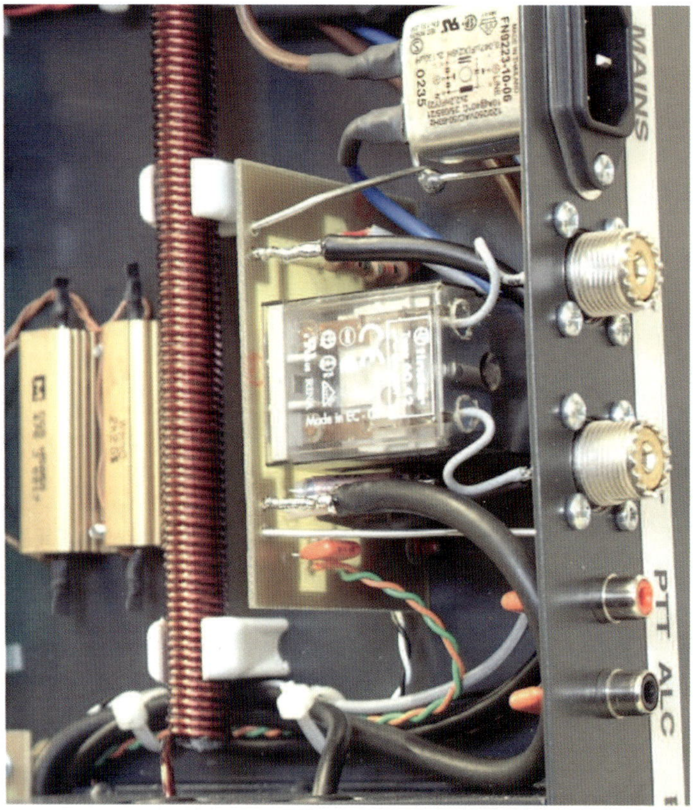

The cathode choke is to the left of the board and wound with 2mm varnished wire on a ferrite rod 12mm x 150 mm.

The mains input filter is in the top right of the picture

The coax bypass circuit is a 20 amp double pole relay. The 2 pole wires are removed from their base pins and the 2 holes are drilled in the back of the relay so they can be wired directly to the SO239 sockets.

Anode Choke CH6

20mm PTFE former 5in /120mm long, ceramic, fibre glass or Teflon will also be fine.
110 turns of 1mm enamelled wire.

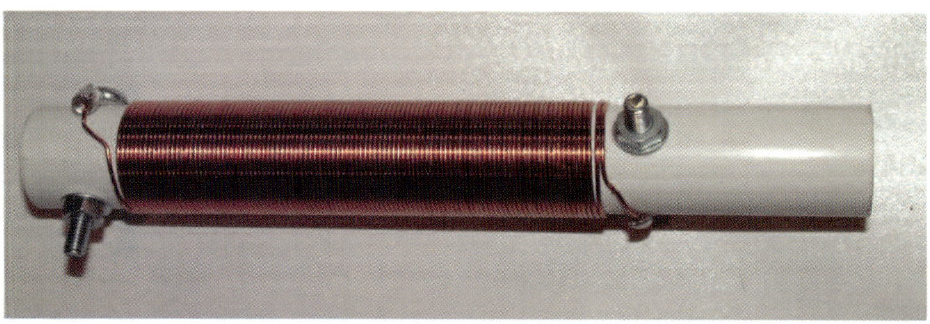

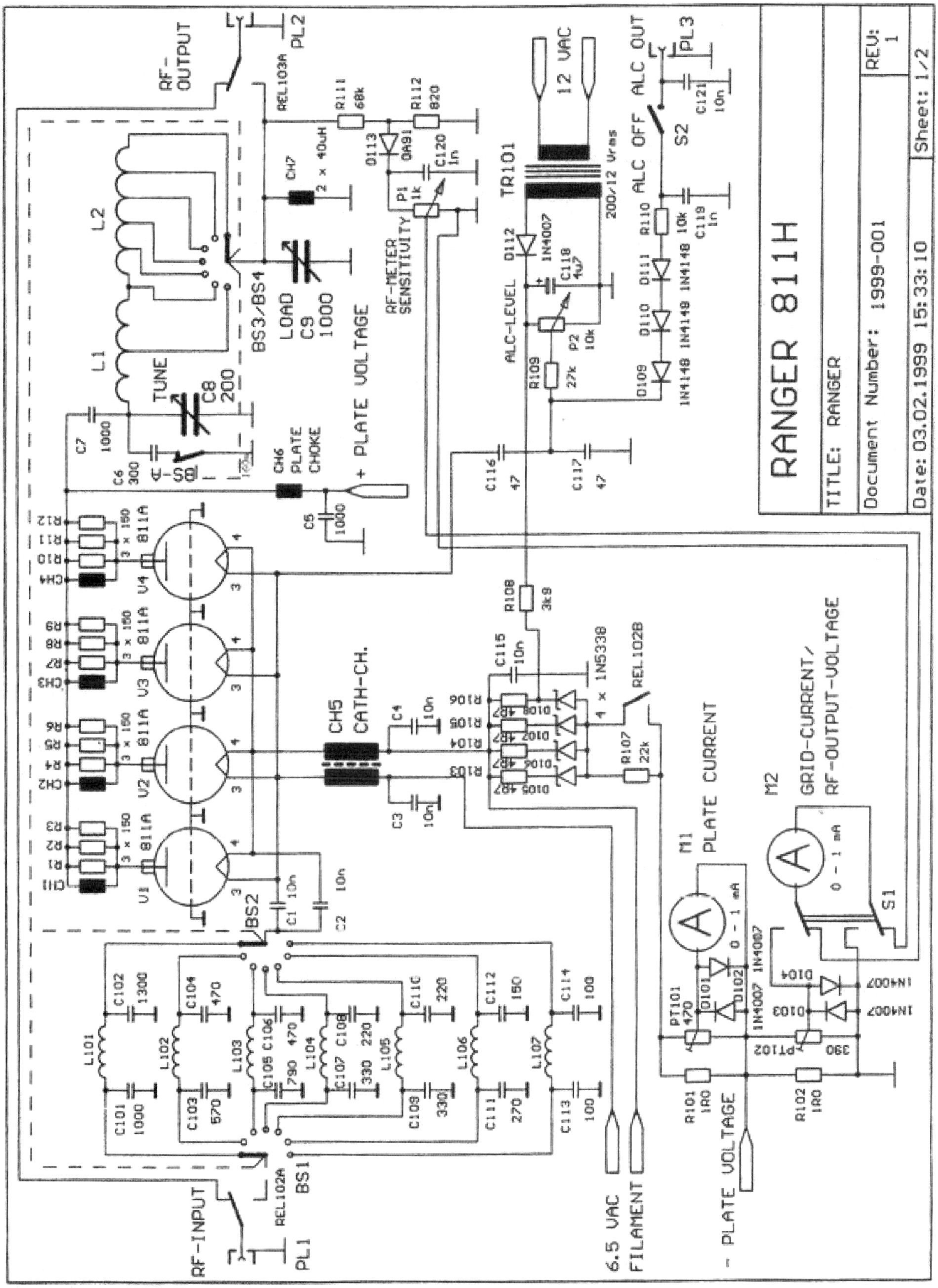

RF circuit of the 811 amplifier

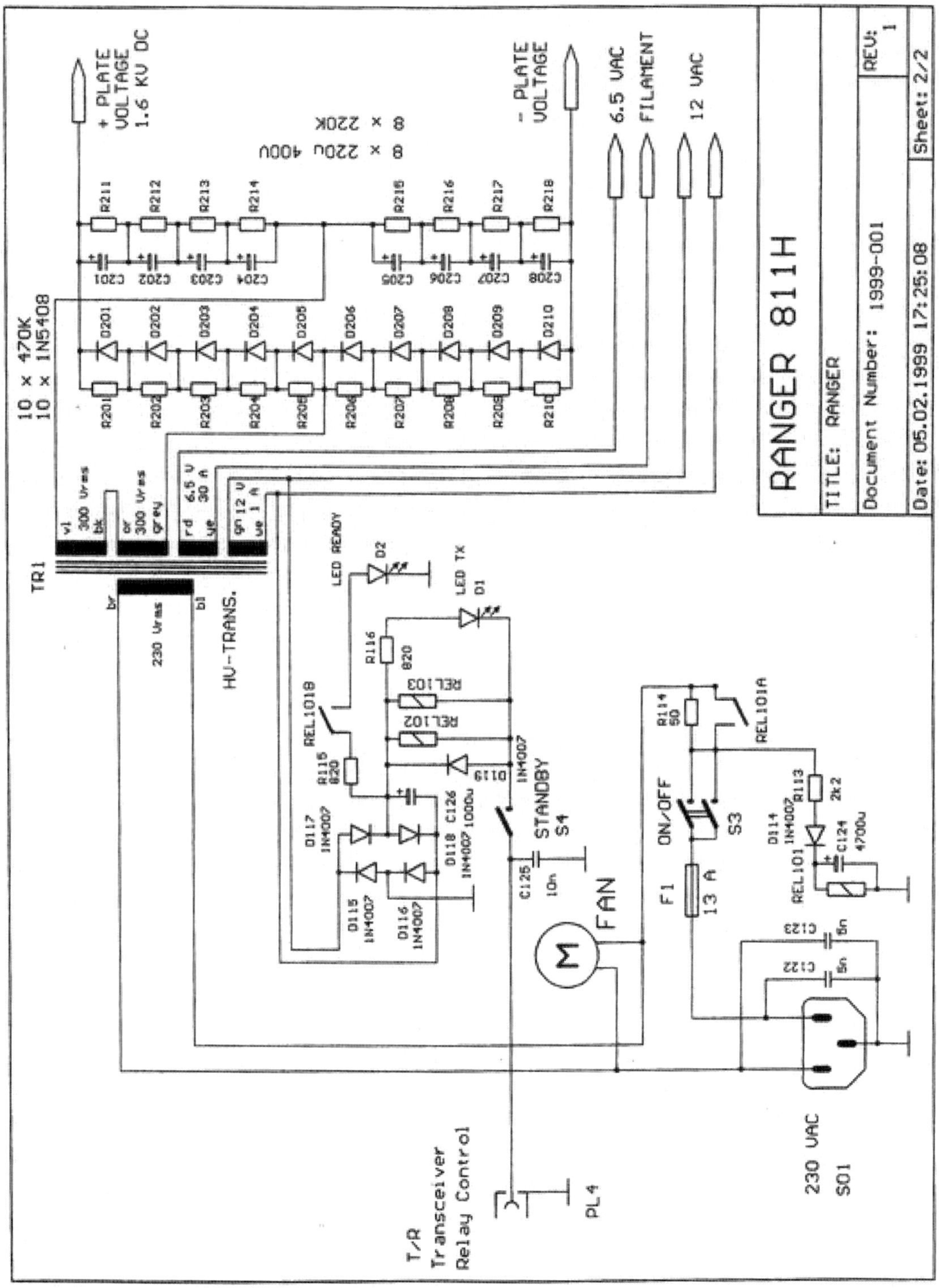

Power supply unit and control circuitry of the 811 amplifier

RANGER 811H

Partslist of electronic components

Ref.: Document Number 1999-001

Pos7	Qty	Designation	Type/Value
001	1	BS1	bandswitch / special ceramic switch
002	1	BS2	bandswitch / special ceramic switch
003	1	BS-A	160m-sw / LA UK
004	1	BS3/BS4	bandswitch / special ceramic switch
005	4	C1, C2, C3, C4	cer. cap. / 10n, 400 V
006	2	C5, C7	cer. cap / 1000p, 10 KV
007	1	C6	H.E.C. / 300p, 7.5 KV
008	1	C8	var. cap. LA UK / 200p, 6 KV, air
009	1	C9	var. cap. LA UK / 1000p, 3 KV, air
010	4	CH1 - CH4	supp. choke / LA UK
011	1	CH5	cathode-choke / LA UK
012	1	CH6	plate-choke / LA UK
013	1	CH7	static supp. / 2 x 40 uH
014	1	D1	LED TX
015	1	D2	LED READY
016	1	F1	Fuse / 13 A
017	1	FAN	Fan / Pabst 4850 N
018	1	L1	Inductor / LA UK
019	1	L2	Inductor / Annecke 28 uH
020	2	M1, M2	Amp-Meter / 0 - 1 mA, SIFAM
021	1	P1	Potentiometer / 1k, 0.5 W
022	1	P2	Potentiometer / 10k, 0.5 W
023	2	PL1, PL2	Coax-Conn. / SO239, Amphenol
024	2	PL3, PL4	Phono-Conn. / RCA phono (Cinch)
025	12	R1 - R12	Resistor / 220, 2 W
026	1	S1	Switch / DPDT, Meter Grid C./RF-Volt.
027	1	S2	" / SPST, ALC Off
028	1	S3	" / DPST, Mains On
029	1	S4	" / SPST, Standby
030	1	SO1	Mains Conn. / 230 VAC
031	1	TR1	Main Transformer / LA UK, 1 KVA Toroid
032	4	V1 - V4	Valve / 811A, Svetlana
101	1	C101	cer. cap. / 1000p, 2 KV
102	1	C102	" / 1300p, 2 KV
103	1	C103	" / 570p, 2 KV
104	1	C104	" / 470p, 2 KV
105	1	C105	" / 790p, 2 KV
106	1	C106	" / 470p, 2 KV
107	1	C107	" / 330p, 2 KV
108	1	C108	" / 220p, 2 KV
109	1	C109	" / 330p, 2 KV
110	1	C110	" / 220p, 2 KV
111	1	C111	" / 270p, 2 KV
112	1	C112	" / 150p, 2 KV
113	1	C113	" / 100p, 2 KV
114	1	C114	" / 100p, 2 KV
115	1	C115	" / 10n, 400 V
116	2	C116, C117	" / 47p, 200 V
117	1	C118	el. cap. / 4.7u, 400 V
118	2	C119, C120	cer. cap. / 1n, 400 V
119	1	C121	" / 10n, 400 V
120	2	C122, C123	foil cap. / 5 n, 230 VAC
121	1	C124	el. cap. / 4700u, 40 V
122	1	C125	cer. cap. / 10 n, 400 V
123	1	C126	el. cap. / 1000u, 25 V
124	4	D101 - D104	Si-Diode / 1N4007
125	4	D105 - D108	Zenerdiode / 1N5338
126	3	D109 - D110	Si-Diode / 1N4148
127	1	D113	Ge-Diode / OA91
128	7	D112, D114 - D119	Si-Diode / 1N4007
129	7	L101 - L107	Inductor Input / LA UK
130	1	PT101	Trimm-Pot. / 470, 0.5 W
131	1	PT102	" / 390, 0.5 W
132	2	R101, R102	Cer. Resistor / 1R0, 4 W
133	4	R103 - 106	Resistor / 4R7, 4 W
134	1	R107	" / 22k, 7 W
135	1	R108	" / 3k9, 0.5 W
136	1	R109	" / 27k, 0.5 W
137	1	R110	" / 10k, 0.5 W
138	1	R111	" / 68k, 0.5 W
139	1	R112	" / 820, 0.5 W
140	1	R113	" / 2k2, 50 W
141	1	R114	" / 50, 50 W
142	2	R115, R116	" / 820, 0.5 W
143	1	REL101	Relay / 24 V, D.Pole
144	1	REL102	" / 12 V, Finsel, D.Pole
145	1	REL103	" / 12 V, D.Pole
146	1	TR101	Transformer 12/230 V
201	8	C201 - C208	el. cap. / 220u, 400 VDC
202	10	D201 - D210	Si-Diode / 1N5408, 1200 V @ 3 A
203	10	R201 - R210	Resistor / 470k, 2 W
204	8	R211 - R218	" / 220k, 2 W

Remark:
Position-Numbers starting with 0 are mounted to the front-panel or the chassis
Position-Numbers starting with 1 are mounted on the main circuit board
Position-Numbers starting with 2 are mounted on the high voltage power supply board

Making a 1000w 2 x 3-500Z HF Amplifier

Introduction

For many years the 3-500Z has been one of the most popular valves used singly or as a pair.

The Drake L4/L7, Kenwood TL922, Heathkit SB220, Henry, Linear Amp UK Explorer 1000 and 1200, and Tokyo Hi Power are few famous makes that have used a pair of 3-500Z.

The Ameritron AL80b, Drake L75 and the Lin Amp Hunter 750 used a single valve to deliver 700 to 800 watts.

The 3-500Z was originally by Eimac in the USA and later by Amperex in France and also Marconi in Chelmsford England under the brand of EEV, now the only manufacturer is the Chinese and re-badged under many names like Taylor and others.

Amperex and the Chinese introduced the graphite anode and changed the name to the 3-500ZG.

The graphite anode is quite an improvement. The valve anode runs cooler at full power compared with the pressed metal anodes.

One of my first HF amplifiers was built from an article in an old ARRL hand book. It used 2 x 4-400 in grounded grid, these valves were readily available as many marine transmitters used a pair to give 1000 watts. They are a tetrode and normally have a passive grid tuned input circuit. They can be driven with a low drive of 10 to 20 watts. Most Amateur transmitters were made with 100 watts so the high gain of a tetrode was not required. The grids can be strapped to make the tetrode into a triode and depending on the HT voltage a pair would give anything between 800 watts and 1200 watts.

3-500Z are a very stable valve and work well up to 30 MHz and can be used up to 50 MHz with careful circuit design.

Being a triode it only requires a single HT supply hence giving good reliability. Thousands are in use around the world every day. The 3-500Z does not need a complicated control board, none of the older manufacturers added any extra protection other than safety fuses and maybe some inrush protection.

The described circuit will work with the 3-500Z, 3-500ZG or 4-250 and 4-400

My first production 1000 watt amplifier was the Explorer 1000 which is similar to the Heathkit SB220 and Kenwood TL922.

The front panel layout of the Explorer 1200, it has all bands including the WARC bands, fitted with soft start to prolong the life of the heaters. The layout is simple and easy to copy.

Top view of the 3-500Z amplifier

The case is size

410mm wide x 245mm high x 450mm deep

This model runs from a single 1200VA toroidal transformer which can be seen in the top right.
The HT power supply board is mounted on the right and it has 1000 volts AC going to it delivering 2800 volts DC.

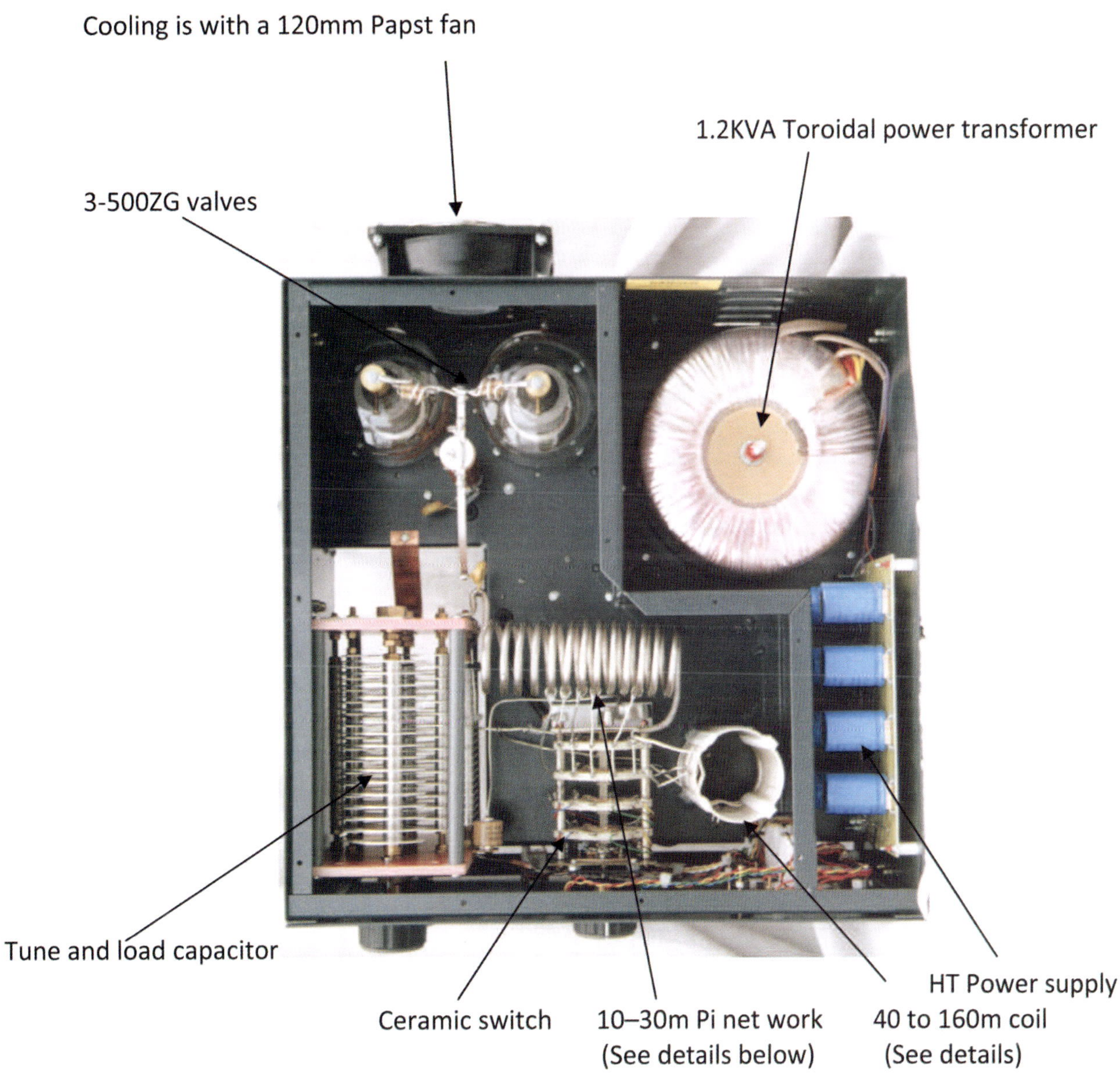

PA Coil

10m to 30m

The 10 to 30m PA coil is made of 6mm copper tube. You need a former which is 38mm or 1.5 inch in diameter. It is nice to get the coil silver plated if you have someone local who can do it. This is not necessary but purely cosmetic and looks nice.

To fasten the connections on to the coil, use the idea below, but you must use a large soldering iron in the region of 250 watts or a gas soldering iron.

The taps on the coil are as follows

Band	10m	12m	15 m	17m	20m	30 m	
Turns	4	5	6	7	10	13	(connected to the top of L2)

40m to 160m coil details, mounted vertically on base of amplifier

L2 is for 40, 80 and 160m and can be made of B & W coil stock 2 inch /50mm, 1.5mm silvered copper wire. A total of 26 turns and tapped from the top at:-

Turns	
0	to the end L1
6	40m connection
18	80m connection
26	160 connected to pole and last contact of BS3

Alternatively L2
Made from 24 turns close wound of 1.5mm varnished copper wire.
Wound on a fibre glass former of 1.5 inch/ 40mm diameter.

Turns	
0	to the end of L1
6	40 m connection
17	80m connection
24	160m connected to pole and last contact of BS3

APC

Wind a 2 turn coil made of 2 or 3mm silvered wire on a 12mm former.
Join together 3 x 150 ohm 2 watt carbon resistors in parallel, see picture of top of valves
The resistors are then placed inside the coil and soldered at each end.
The APC (anti parasitic choke), or rather the L in the APC forms a choke that stops anything radiating above 30 MHz

If the amplifier is used to its very highest frequency such as 29.6 FM the APC could get close to its cut off frequency. If this occurs the 3 x 150 ohm resistors will get hot or even burn out.
The APC can be stretched a certain amount, which increases the cut off frequency and allows operation up to 29.6 MHz

Cathode Choke

Wind on a 12mm former with 2 x 2.5 mm varnished copper wire, then insert a 12mm ferrite rod.
Wind first on a wood dowel of 12mm or you will crack the ferrite whilst winding the heavy wire.

Layout of the underside of the 2x 3-500 amplifier

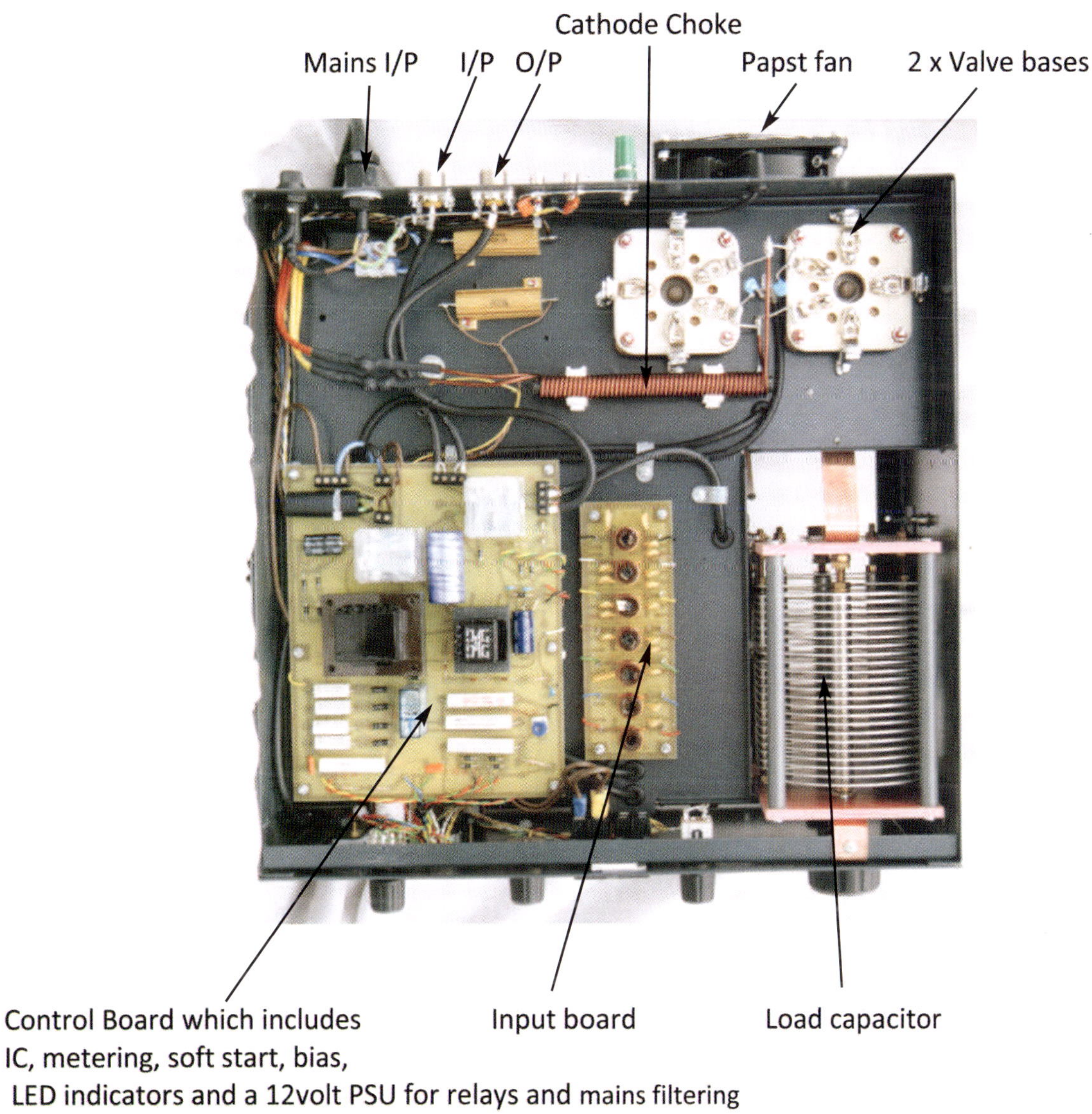

This layout gives good isolation between the input circuits and the output. The valve bases are mounted over a large hole to allow plenty of air to circulate around the valve bases. Good cooling is very important to the valve base. You cannot have too much cooling. A 3-500Z gets very hot around the heater pins and must be kept cool.

3-500Z valve bases

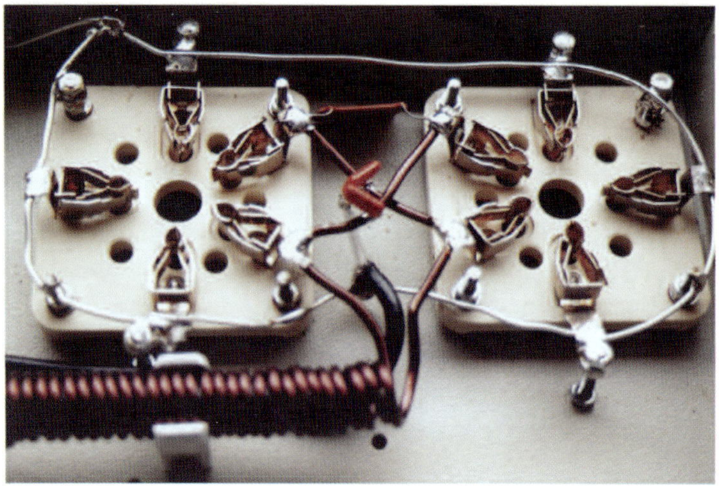

Above is a close up of the valve bases. All the grid pins are linked together and earthed. The heater pins are 1 and 5 and are fed from the end of the Cathode choke. Some amplifiers use a grid choke to earth from the grids. Eimac recommend direct earth connection. It is up to you which you choose.

Biasing

Biasing 3-500Z needs +5 volts of bias on the cathode at 2800 volts. This is done by almost all manufacturers by using a zener diode. A BZY93 is a good choice or 4 x 5 watt parallel 5 volts zener diodes will also work fine as can be seen on the control board above.

Input board for 2 x 3-500Z

OUTPUT to cathode

Band	10/12m	15/17m	20m	30m	40m	80m	160m
O/P capacitor	0pF	0pF	150pF	220pF	320pF	690pF	2200pf

I/P capacitor	100pF	100pF	220pF	320pF	470pF	690pF	1700pF

INPUT from band selector

There is no capacitance on the output for 10/12m and 15/17m as there is adequate capacitance between the valve cathodes to grid.

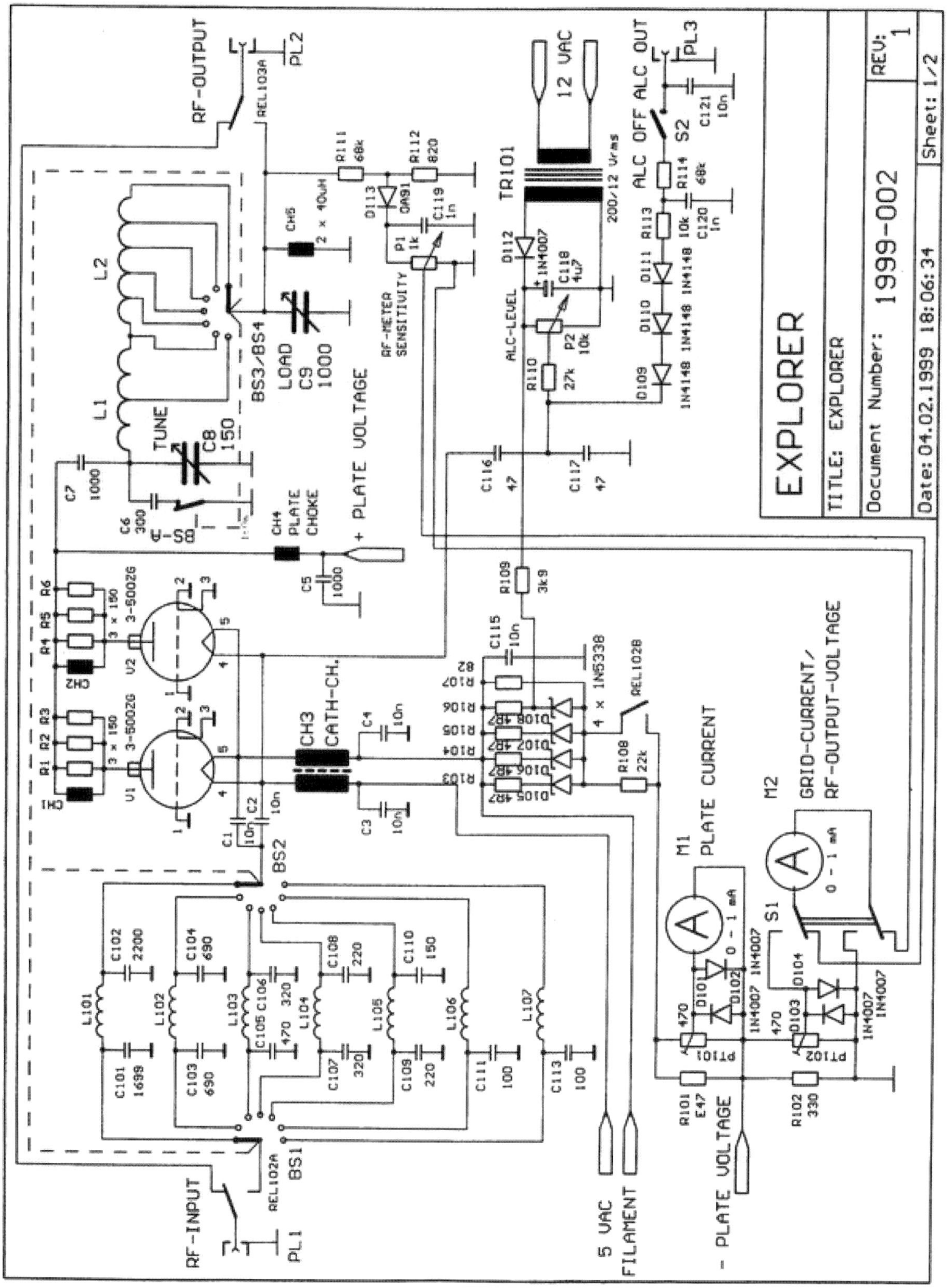
EXPLORER
TITLE: EXPLORER
Document Number: 1999-002
Date: 04.02.1999 18:06:34
REV: 1
Sheet: 1/2
RF-OUTPUT
PL2
RF-INPUT
PL1
PL3
ALC OFF ALC OUT
12 VAC
5 VAC FILAMENT
- PLATE VOLTAGE
+ PLATE VOLTAGE
PLATE CURRENT
GRID-CURRENT/ RF-OUTPUT-VOLTAGE
TUNE
LOAD
CH3 CATH-CH.
CH4 PLATE CHOKE
RF-METER SENSITIVITY
ALC-LEVEL

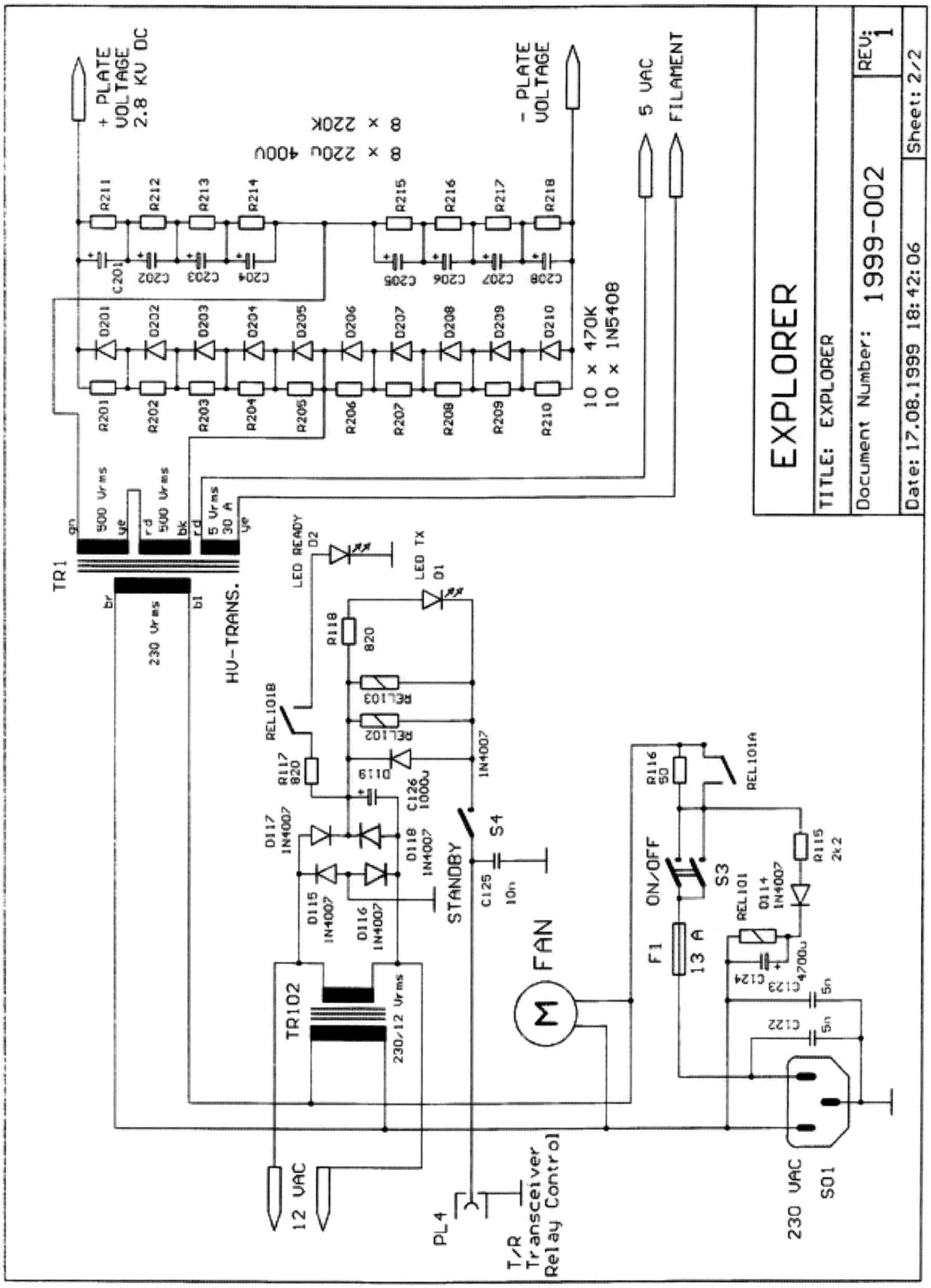
+ PLATE
VOLTAGE
2.8 KV DC
- PLATE
VOLTAGE
5 VAC
FILAMENT
8 x 220K
8 x 220u 400V
10 x 470K
10 x 1N5408
TR1
HV-TRANS.
230 Vrms
500 Vrms
500 Vrms
5 Vrms
30 A
LED READY
D2
LED TX
D1
R118
820
REL103
REL102
REL101B
R117
820
D119
C126
1000u
1N4007
D117
1N4007
D118
1N4007
D115
1N4007
D116
1N4007
S4
STANDBY
C125
10n
TR102
230/12 Vrms
FAN
F1
13 A
ON/OFF
S3
R116
50
REL101A
R115
2k2
REL101
D114
1N4007
C124
C123
4700u
C122
5n
5n
12 VAC
PL4
T/R
Transceiver
Relay Control
230 VAC
SO1
EXPLORER
TITLE: EXPLORER
Document Number: 1999-002
REV: 1
Date: 17.08.1999 18:42:06
Sheet: 2/2

Making a 1500 watt Ceramic Triode Amplifier

This is the last triode amplifier I am going to describe and is the last of the Challenger amplifiers. It is still manufactured by Roger Banks at the DXshop in Wales. It can be bought complete or the parts including cabinet valves and transformers etc can be bought separately.
It started out as the Challenger I using 2 x 3CX800A7 as shown below.

This was my first venture into ceramic triodes and, wow, I was so impressed with them, and still am.

I changed to the 3CX1500A7 and finally ended up using the Russian GS35b. The largest cost of any amplifier is the valve and this determines which option I used for each Challenger.

The amplifier described will work with either the GS35b or the 3CX1500A7.

Both these valves will happily give 1500 watts but the cost of the GS35b beats the other. If you would like an amplifier that will give 1200 watts with an HT of 2800 volts use the GS31b.

Most glass valves have a gain of around 10 but if you want to get to the 1500 watt level with a 80 to 100 watt driver you will need something with a bit more gain. This can be done with a tetrode or the modern hi μ triode which has a gain of around 40.

The main change from glass valves is the air has to be directed through the external anode cooler. One small 3CX800A7 will almost produce as much power as 2 x 3-500Z. The valve still has to dissipate the heat off the anode but this is done with much more efficiency than using a glass envelope valve.

All ceramic triodes are indirectly heated (except the 3CX1200 which is more like a high power 3-500Z and uses the 3-500Z valve base).

Though not compulsory it is a very good idea to fit grid and anode current limiting. The most delicate part of a ceramic tube valve is the grid. A 3 min warm up timer should be used to stop the danger of stripping the cathode.

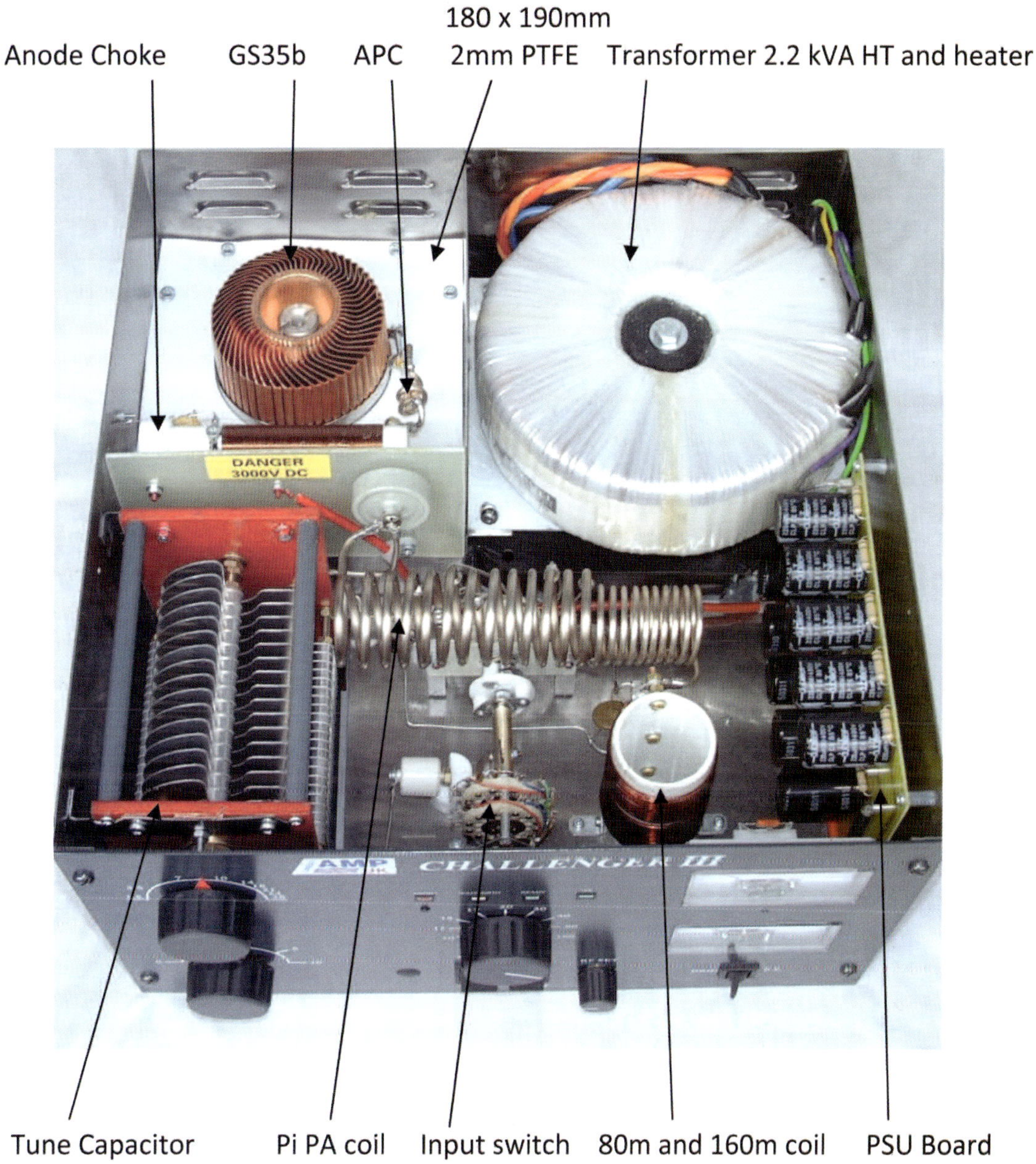

The cabinet size is 400mm wide x 230mm high x 420mm deep
The cabinet can be made of aluminium or steel, this one is steel, the transformer and valve are heavy so adequate support must be considered.

The cooling fan is a Papst flatpack blower, which is mounted underneath the mains transformer. Before the air is blown into the compartment under the valve the air passes around the mains transformer and helps keep the whole cabinet cool. The air exhausts out above the valve.

Tuned input board

Band	10/12m	15/17m	20m	30m	40m	80m	160m
Input capacitor pF	100	150	220	330	470	1470	2000
Number of turns	4	5	6	8	9	16	27
Output capacitor pF	0	100	150	220	470	1000	2000

Note 10-20m coils are wound with 1mm varnished copper wire
30-160m coils are wound with 0.5mm varnished copper wire

Anode Choke

135mm x 20mm PTFE
Wound with 0.8mm varnished copper wire, 110 turns.
The fixing screws are 100mm apart. Use a clear lacquer to varnish all turns when finished.

APC (anti parasitic choke)

2 x turns of 3mm copper tube, silver plated
3 x 150 ohm 2 watt carbon resistors in parallel mounted on the inside of the coil

Connection on the GS35b

The base of the GS35b is secured straight into a 60mm hole in the amplifier chassis, the valve is clamped into the hole with 25mm x 25mm wide L brackets.

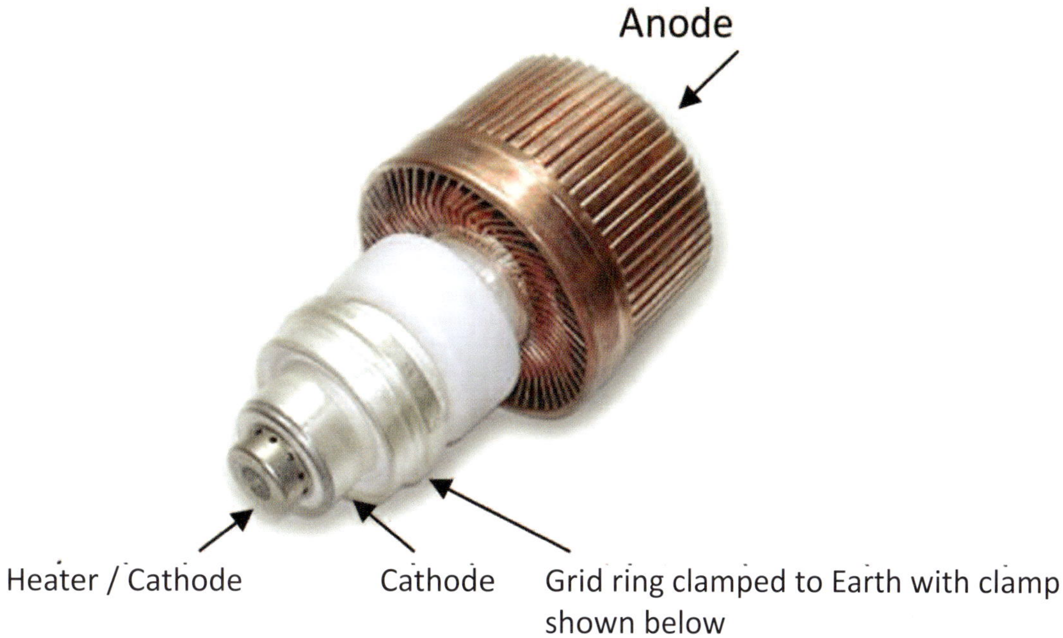

Clamp 25mmx 25mm and 7mm high x 4 used to hold valve into chassis.

Close-up view of the PA Pi tank circuit

Tune capacitor
Load capacitor
extra 300pF for 160m and switch
10-30m L4a
30-40m L4b
80-160m L5

The picture shows the two pole input switch at the front of the amplifier. The output switch is sited behind and below the PA coil and connected to the I/P switch by a brass shaft. The leaf switch to bring in the extra C for 160m is fastened to this shaft.

PA- Pi Coil winding details

10m – 30m and 30m to 40m (L4a and L4b) and 80m to 160m (L5)

L4a is 13 turns of 6mm copper tube wound on a 1.5 inch / 38mm former
Silver plated or can be cleaned and varnished.

L4b is 10 turns of 3mm copper tube on a 1.25inch/30mm former which is fastened on the end of L4a inside and soldered. This can also be silverplated or cleaned and varnished.

L5 is 30 turns of 1.9mm varnished copper wire wound on a 50mm fibre glass former. The end of L4b is fastened to the top of the L5 winding and forms the 40m tap.

L4a	Band		10m	12m	15m	17m	20m	
From Tune capacitor Turns: -		0	4	5	6	7	11	13 turns total
L4b	Band		30m	40m				
From the end of L4a Turns		0	2	10				
L5	Band		80m	160m				
From end of L4b			13	30				

Below is L5 which is the 80m and 160m part of the Pi PA coil

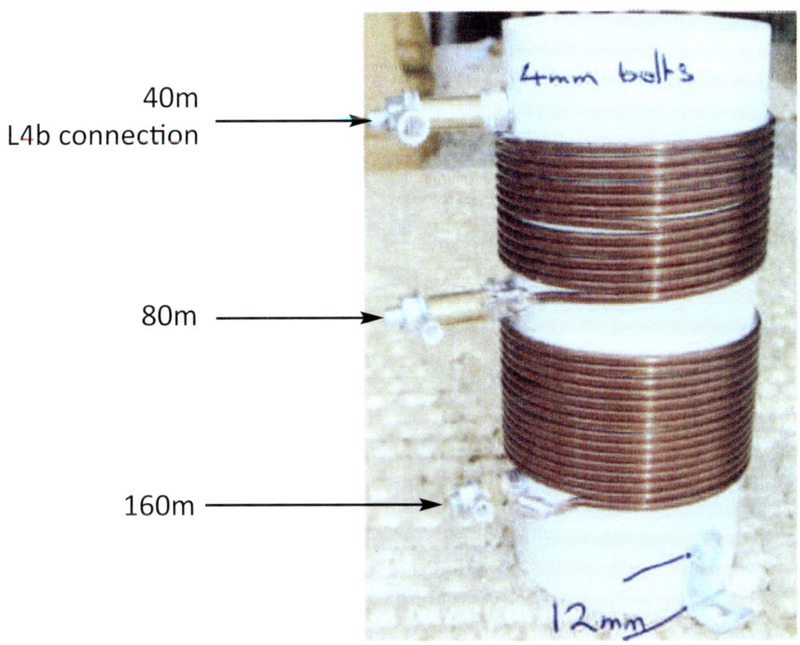

Control Board.

The control circuitry can be hard wired and just do the basics, alternatively I would recommend the GM3SEK Triode board which is available from Ian White, this does everything including all protection, metering and start up timing. It also has a variable bias supply, the cathode voltage on transmit wants to be close to 36 volts. An alternative can be a BZY93 zener diode which I have used with great success for many years.

Drawing 1

RF drawing of GS35 amplifier and the connection details for the GM3SEK board. The hand book that comes with the control board is very easy to follow and you won't have any problem setting it up as all setting up is detailed in the hand book.

Drawing 2

This is my control board which is more relay based. It works well and is still used in all Challengers and Discovery amplifiers made now.

Drawing 3

Power supply board as used in the Challenger 3 and 4.

Drawing 1

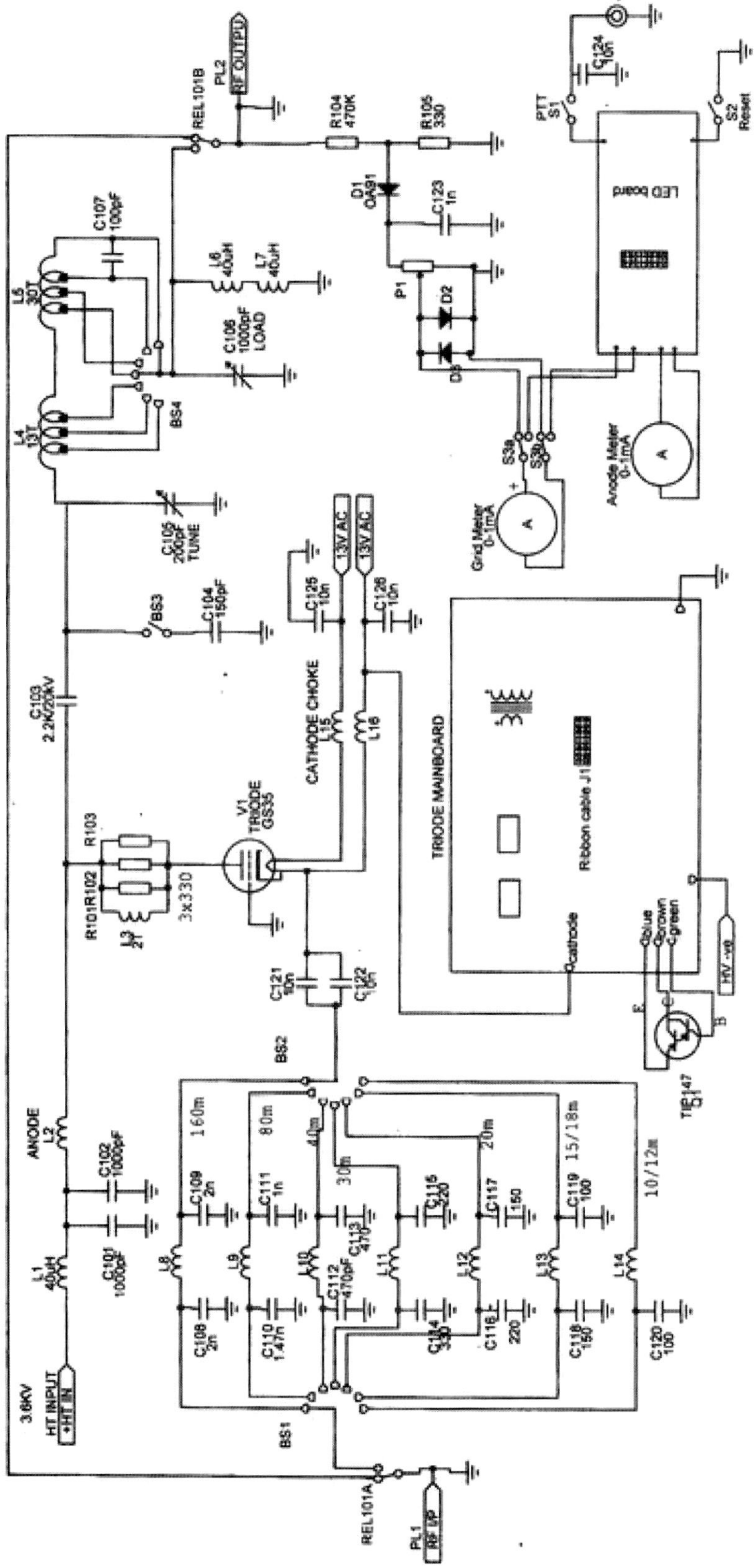

REL101B
PL2
RF OUTPUT
R104
470K
R105
330
J1
PTT
S1
C124
10n
S2
Reset
C107
100pF
L5
30T
L6
40uH
L7
40uH
D1
OA91
C123
1n
LED board
P1
D2
D3
C106
1000pF
LOAD
L4
13T
BS4
S3a
S3b
Anode Meter
0-1mA
Grid Meter
0-1mA
C105
200pF
TUNE
13V AC
13V AC
C125
10n
C126
10n
BS3
C104
150pF
C103
2.2K/20KV
CATHODE CHOKE
L15
L16
TRIODE MAINBOARD
Ribbon cable J1
R103
R101R102
3x330
L3
2T
V1
TRIODE
GS35
cathode
blue
brown
green
HV -ve
C121
10n
C122
10n
E
C
B
TIP147
Q1
BS2
ANODE
L2
C102
1000pF
160m
80m
40m
30m
20m
15/18m
10/12m
C109
2n
C111
1n
C113
470
C115
220
C117
150
C119
100
L1
40uH
C101
1000pF
L8
L9
L10
L11
L12
L13
L14
C108
2n
C110
1.47n
C112
470pF
C114
330
C116
220
C118
150
C120
100
3.6KV
HT INPUT
+HT IN
BS1
REL101A
PL1
RF I/P

Drawing 2

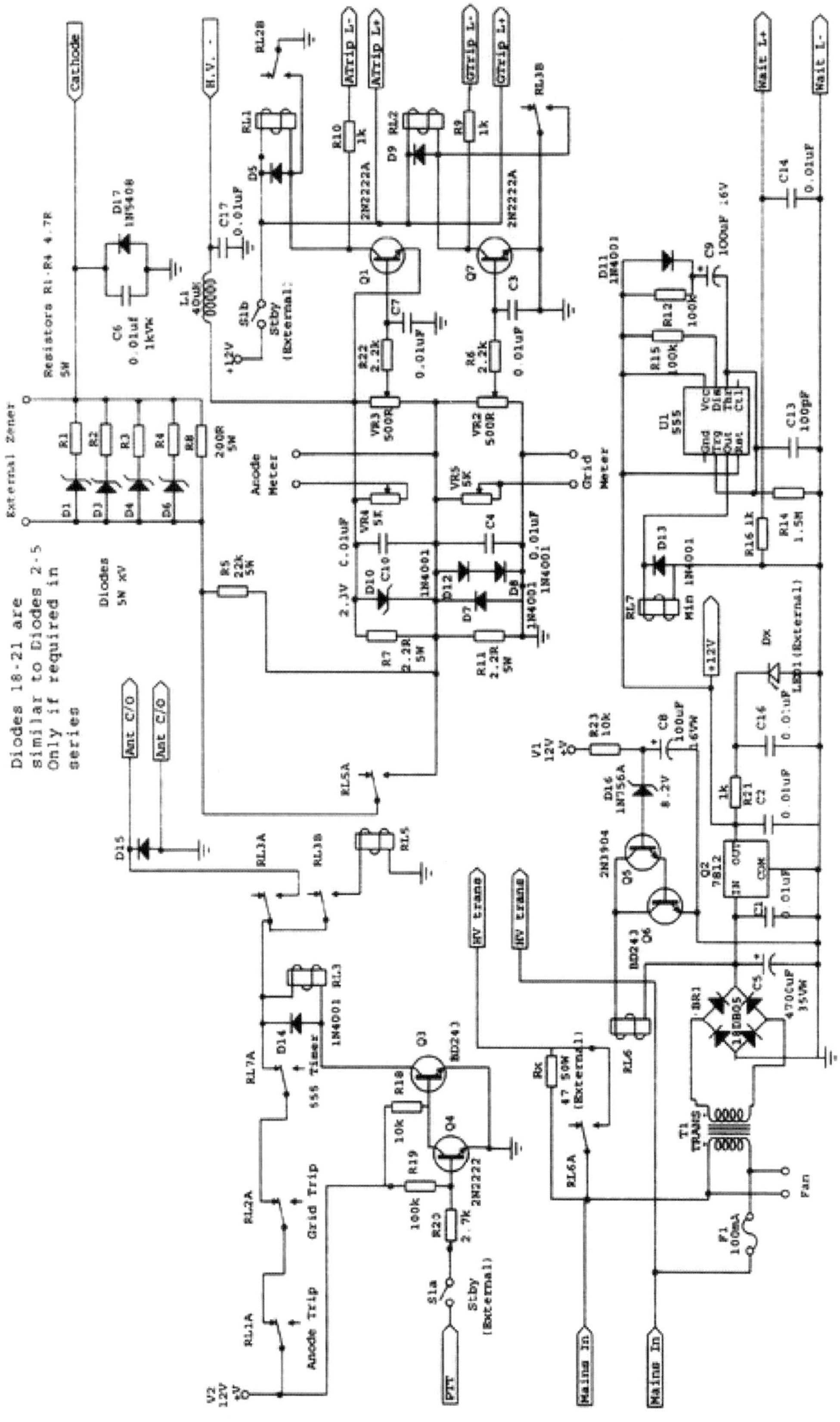

©G3ZRS

Drawing 3
Challenger 3/4 Power supply board

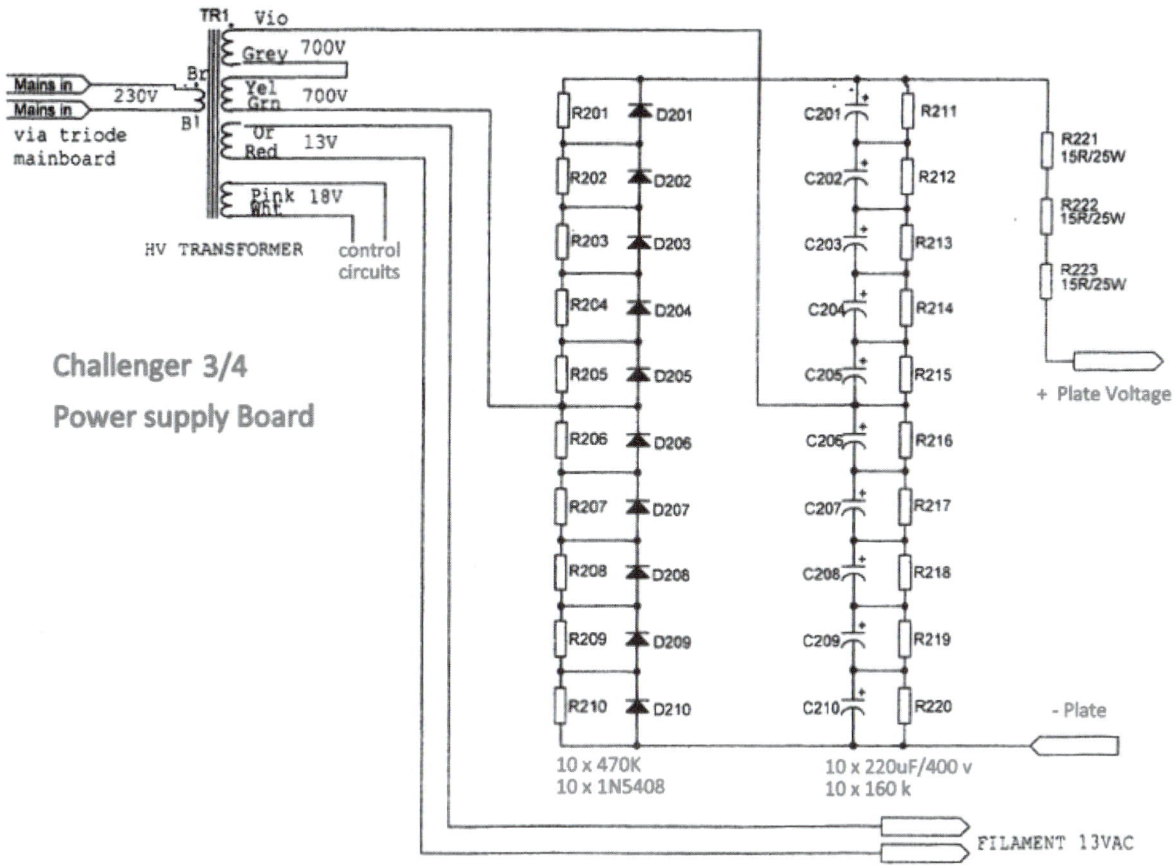

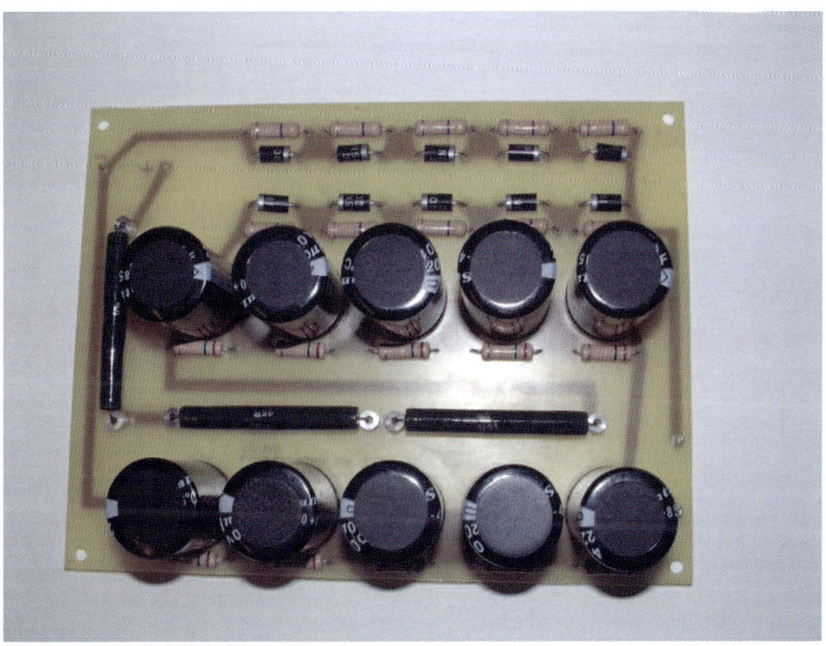

3600 volt power supply Board

Final setting up and tuning

When a triode amplifier has been finished there are a few things that have to be done.

1 The input tuned circuits have to be set and adjusted.

2 Meters need calibrating.

3 Run the amplifier up first with a variac transformer.

1. Tuning the input tuned circuits.

Each band of an amplifier has an individual tuned input circuit which has to be set around its middle frequency.

This can be done with either an antenna analyzer or by using the SWR meter fitted in your radio, or a separate SWR bridge placed between your transmitter and the amplifier input.

If you use the SWR meter in your radio, place it in the CW mode and use about 20 watts of RF.

Tune the input slug for minimum SWR. It should dip to below 1.5 to 1 on each band.

If this cannot be done you must suspect one of capacitors as faulty at one side of the coil to earth.

2. Calibrating your Anode and Grid meter.

When your amplifier is finished you will need to calibrate your anode and grid meters.
This is best done with a 12 volt battery like one from a battery drill.
Use a 100 ohm 2 watt potentiometer and a multimeter capable of measuring 1 amp.

Place the 12 volt battery, 100 ohm pot, and current meter in series across the meter shunt to be calibrated.

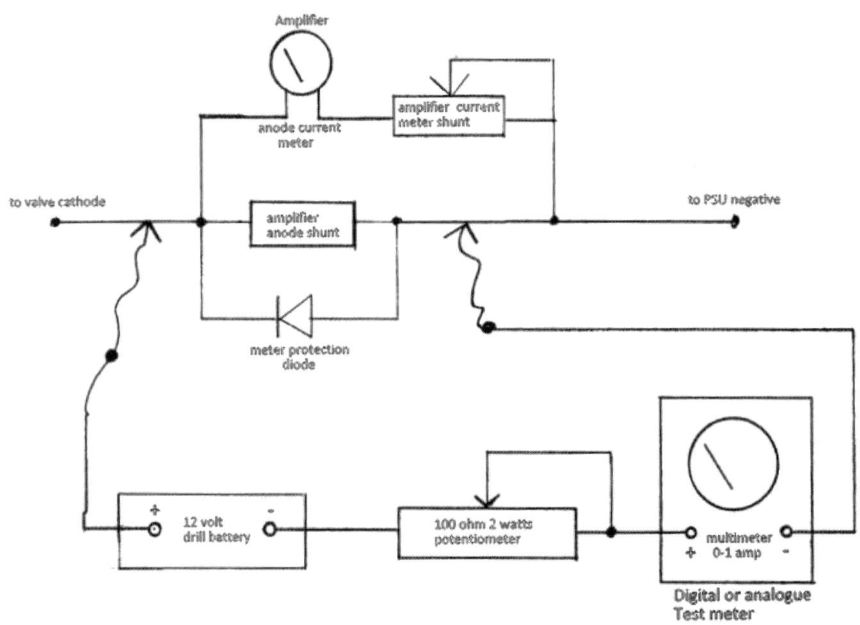

The above circuit can be used on your anode or grid meter. The value of the meter shunts are normally around 1 ohm.

You can repeat the same across the grid shunt which will go from the PSU negative supply to earth.

The multimeter will only need setting to 100mA or 500mA depending on the valve being used.

CAUTION

Final switch on time

Do not run the amplifier when there is any danger of you touching any high voltages.

1 Disconnect one AC wire from the transformer going to the PSU board.

2 You can now try all the other functions of the amplifier with no HT voltage

3 With no HT supply you can still test the input SWR of your amplifier on each band.

4 Finally re-connect the transformer wire going to the PSU board.

If you have access to a variac transformer, use it to run the amplifier up first. You could also fit a low level fuse like 5 amps in the mains supply. Running up the amplifier until you key the amplifier with a variac will not take any more than 5 amps. This will show any faulty wiring and save severe damage.

5 Apply a low level of RF in the CW mode on the 40m band

40m is a good band to test an amplifier for the first time, both the Tune and Load capacitors should be around their mid positions and tune for maximum output on your external power meter. Do not over drive the amplifier.

Ceramic valves often need a running in time. This should be a few hours of heaters only before applying the DC HT supply. This makes sure the vacuum is 100%.

Overdrive of a ceramic valve will cause severe high grid current, adjust your drive till the grid current drops. If your load control is set too low you will also get high grid current, increasing the load control will reduce the grid current back to its allowed current level.

Finally, have fun and always work safely

Chapter 12

Control and Biasing

Control and Biasing

Every amplifier has to have some control circuitry and protection. It can be as simple or as complicated as you wish to make it work. Usually the more modern triode and tetrodes need more protection than older glass valves which just need a bit of current limiting and a mains fuse. Due to their higher gain, ceramic valves need current limiting and a start up timer to ensure the valve is warm enough before putting into operation.

Biasing

A few valves run at zero bias and do not need a bias circuit. These are usually older glass valves running at a low HT voltage, when the HT voltage is raised they do not need any more than 5 to 8 volts of bias which can easily be achieved with a simple zener diode on the cathode.

Triodes

Triodes have a cathode voltage when switched on. If the voltage is left high there will be no current flow. If the voltage is brought down to zero the valve will be fully switched on and the maximum standing current will flow. A controlled stepping of the cathode voltage varies the standing current of the valve. This is not RF but just a flow of electrons like leaving your water tap a quarter open.

Biasing can be done with either a zener diode or a regulator transistor like the TIP 147 that is used on the Triode board from Ian White GM3SEK.

Tetrodes

Tetrodes need their standing current set to put them into the required mode. This is normally done using a negative voltage on grid 1 to control the valve current. No negative voltage on the grid and the valve will take unlimited maximum current. A high negative voltage in the order of -100 volts will bias the valve completely off. A controlled voltage, halfway between zero and maximum voltage will set a standing current to put the valve into AB1 or AB2 for SSB operation. Zero standing current can be used for CW and a higher efficiency can be achieved.

Grid 2

This is commonly known as the screen. A voltage of usually around 10 to 20% of the anode voltage is applied to G2, this can be used to switch the valve on and off in conjunction with the change of bias voltage on Grid1.

If for any reason the anode voltage fails, G2 will think it is the anode and a severe over-current will occur, this must be protected against or the valve screen will be damaged.

Basic control
The minimum control of an amplifier:-

1. A circuit that allows you to receive through the amplifier and transmit through with your radio only. This can be done with a double pole relay, and in more sophisticated amplifiers which have QSK (receive through CW operation) is done with fast vacuum relays or high power PIN diodes.
2. Protection, in the form of current limiting of the anode and grid circuits.
3. An indication of TX or RX, to show the state of the amplifier operation.
4. Mains power on or off.
5. The switching of the valve from TX to RX done with bias as described on the previous page.

Added control for more sophisticated Amplifiers

Start up timer
More modern valves usually have indirectly heated filaments. They usually take less heater power but require a time of 3 minutes to warm. This is done so the cathode does not get stripped, which would decrease the life of valve!

Ceramic valves, both triode and tetrodes have a much higher gain compared with older glass valves due to their focused beam construction. Protection of the anode can be fitted and is usually set to trip around 1 amp

More important
The grid on modern valves is the most delicate and must be protected. Each valve has a different maximum current allowable, for example a 3CX800A7 has only 60mA but can swing much higher when off tune with full drive. Russian triodes like the GS35b can take several hundred milliamps and is not so critical.

Grid 1 on a tetrode is usually where the drive is applied and the bias voltage is set. If a tetrode is badly tuned it can emit power from Grid 1. This must be dumped into a load resistor so the bias voltage cannot climb and upset the stability of the valve.

Heat
Older glass valves are only cooled across their large glass envelope. The cooling is not very efficient as the anode sits in a vacuum so the heat transfer is not good except for radiated heat.

Modern valves have a much more efficient metal external anode cooling. The valve still operates in a vacuum but the anode is extended to a radiator directly connected to the anode itself. Cooling air is passed through a very efficient radiator. Much larger valves can have a cooling jacket and distilled water circulated from a reservoir tank. Smaller valves like the 4CX250B and 3CX800A7 need a fan with a large back pressure to achieve the required amount of air flow at full power.

Over heat sensors are often fitted on most modern ceramic valve amplifiers. This is to protect from melt down if the fan fails or becomes blocked, or just that the amplifier is off tune and creating excessive heat.

Ian White GM3SEK control boards
Many people from all over the world have built an amplifier and employed a GM3SEK control board. They are made for either the Triode or Tetrode valves. These are available from Ian and so look at his excellent web site which gives great detail of every aspect of the control board. It is at very modest cost and I would highly recommend it. These boards are copyright but Ian has kindly allowed me to give details and circuit diagrams of both the triode and tetrode board.

REC-G1-ALC Schematic

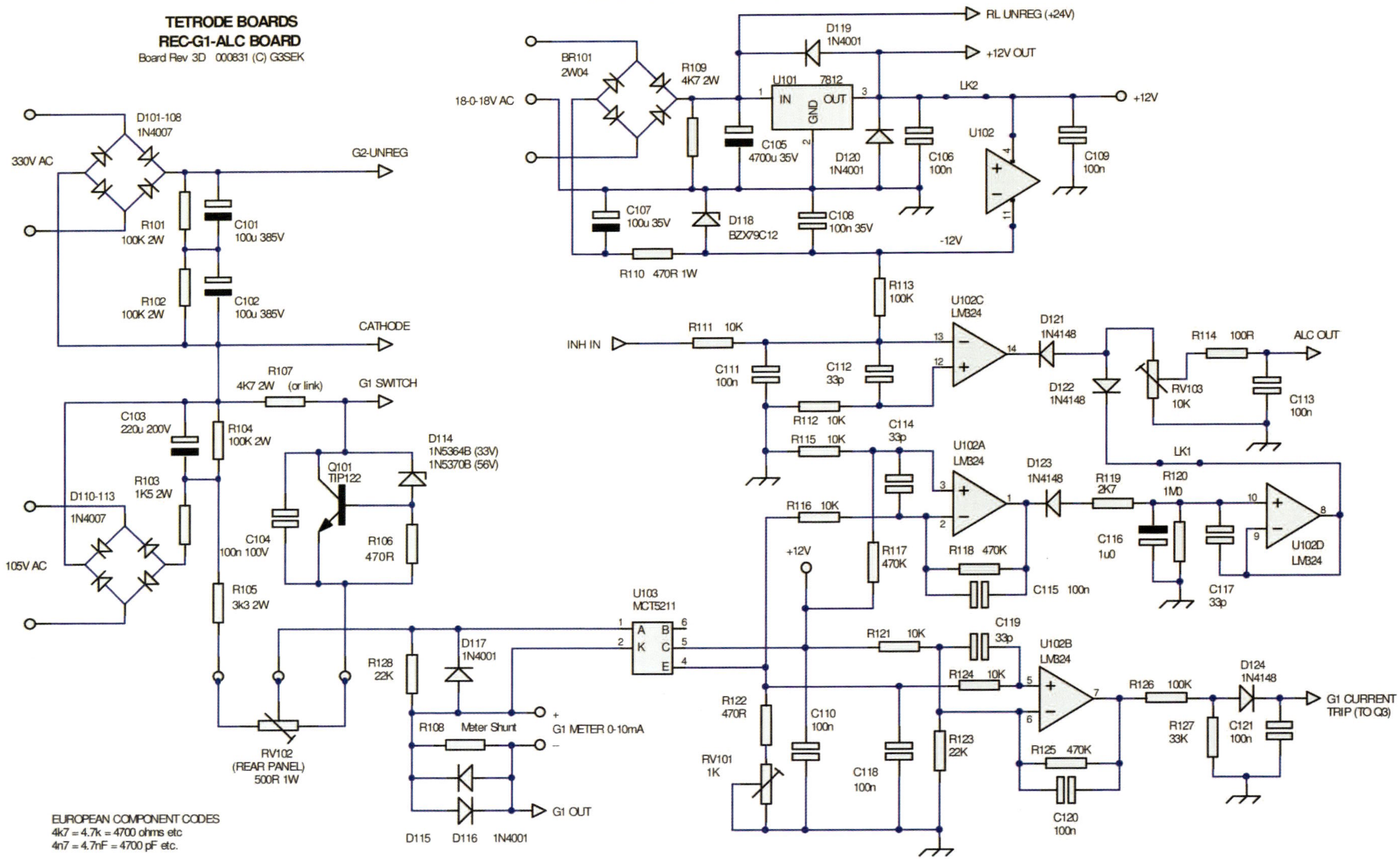

G2-CONTROL Schematic – Sheet 2

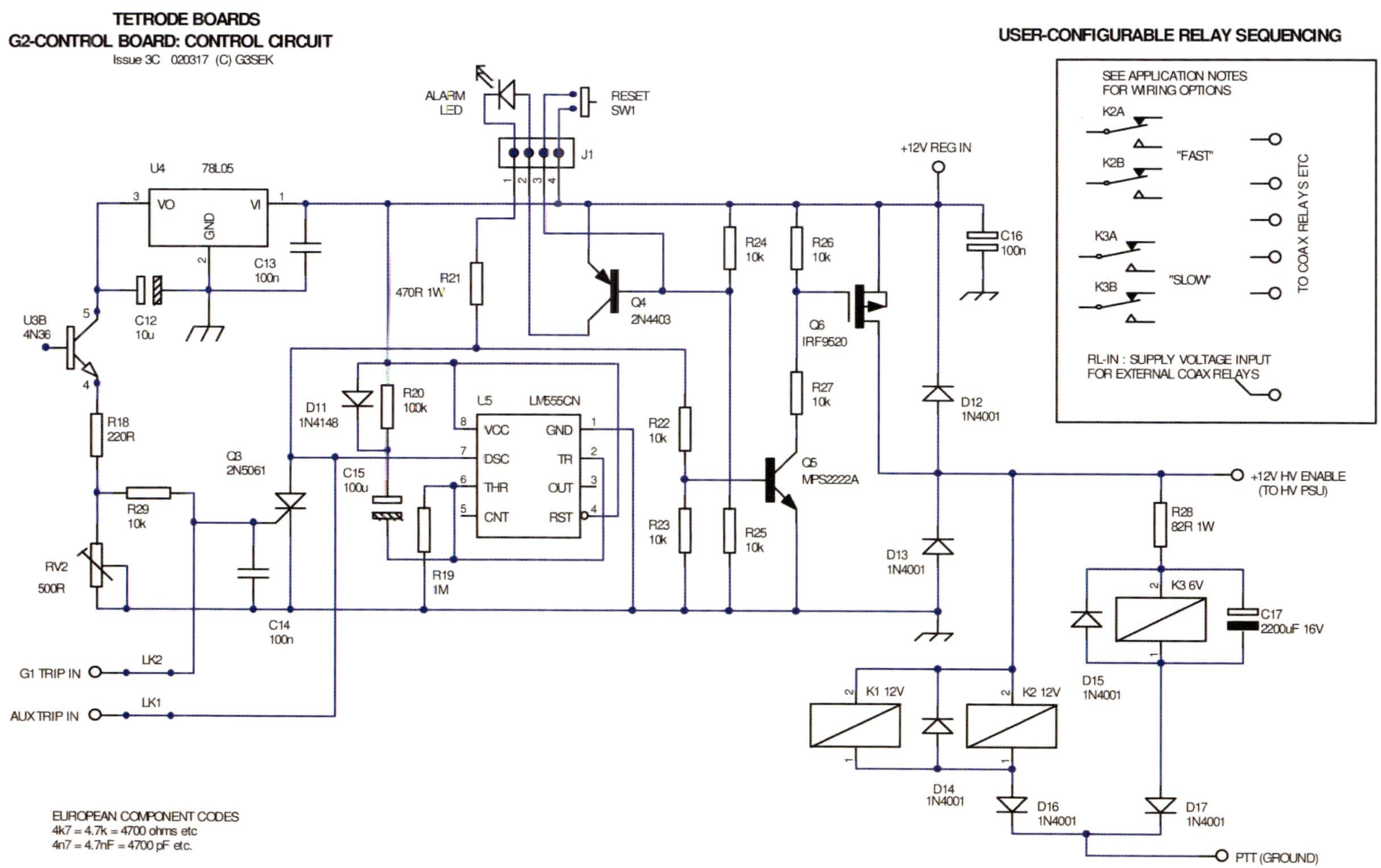

G2-CONTROL Schematic – Sheet 1

TETRODE BOARDS
G2-CONTROL BOARD: SCREEN REGULATOR
Issue 3C 040416 (c) IFWtech Ltd

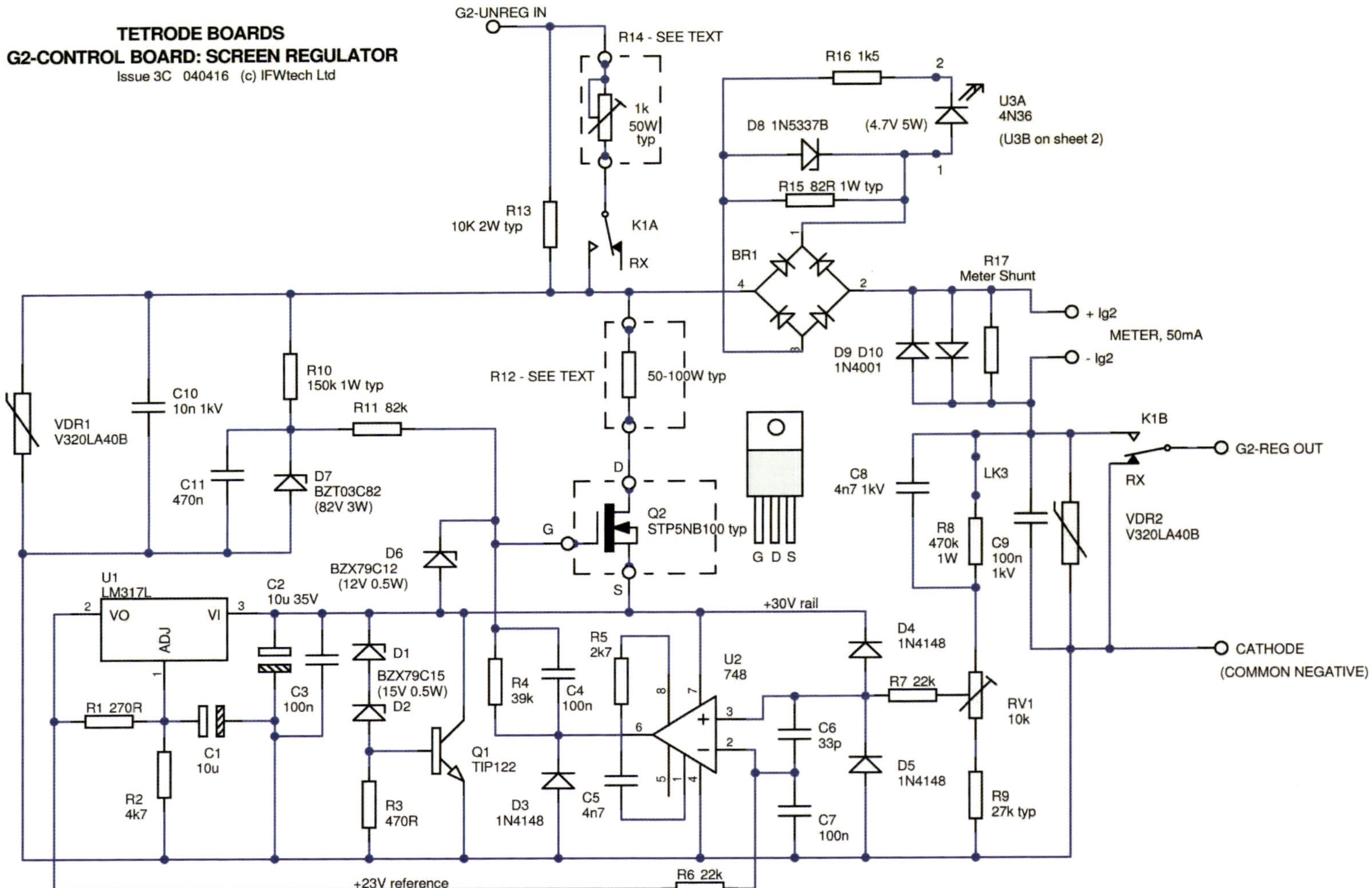

EUROPEAN COMPONENT CODES
4k7 = 4.7k = 4700 ohms etc
4n7 = 4.7nF = 4700 pF etc.

Q1 REQUIRES A SMALL HEATSINK
Q2 REQUIRES A LARGE HEATSINK
R12 AND R14 HAVE HIGH HEAT DISSIPATION

Values marked "Typ" depend on screen voltage and current required - see text and Components List

WARNING
THE SCREEN VOLTAGE INPUT AND OUTPUT ARE UNGROUNDED!
USER MUST PROVIDE GROUNDING AS REQUIRED.

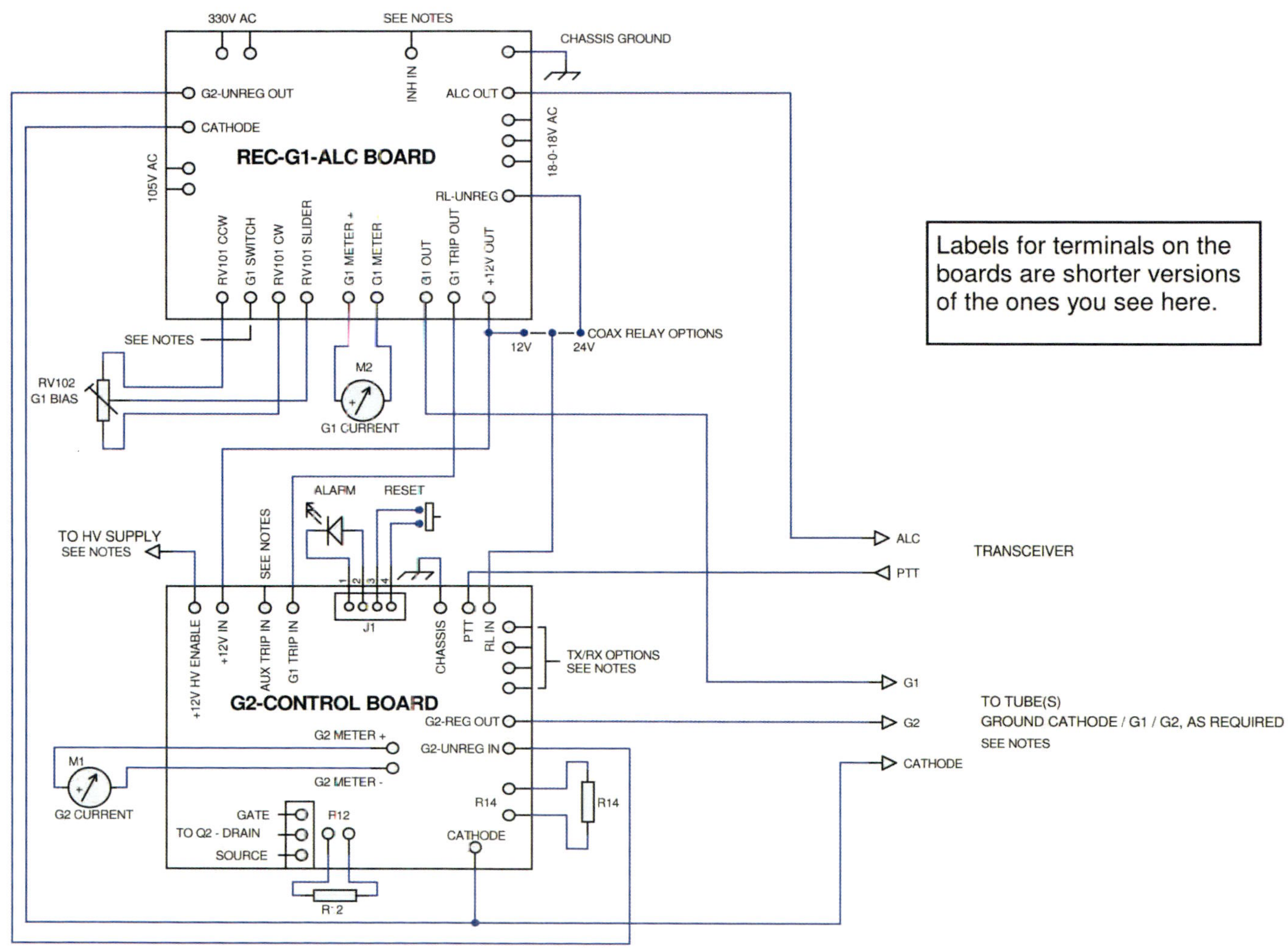

Labels for terminals on the boards are shorter versions of the ones you see here.

The Triode Board Schematic – Sheet 1

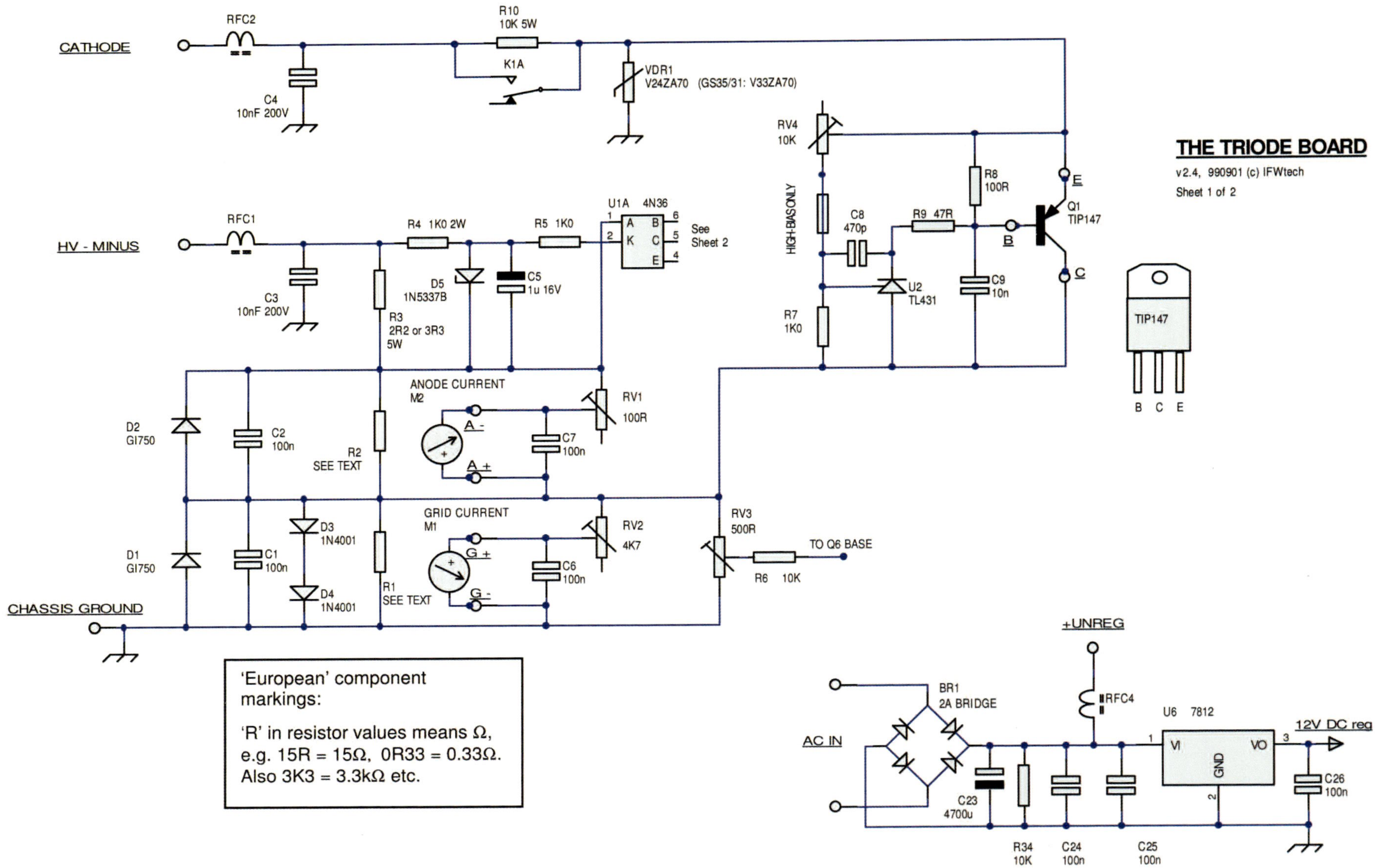

'European' component markings:

'R' in resistor values means Ω, e.g. 15R = 15Ω, 0R33 = 0.33Ω. Also 3K3 = 3.3kΩ etc.

The Triode Board Schematic – Sheet 2

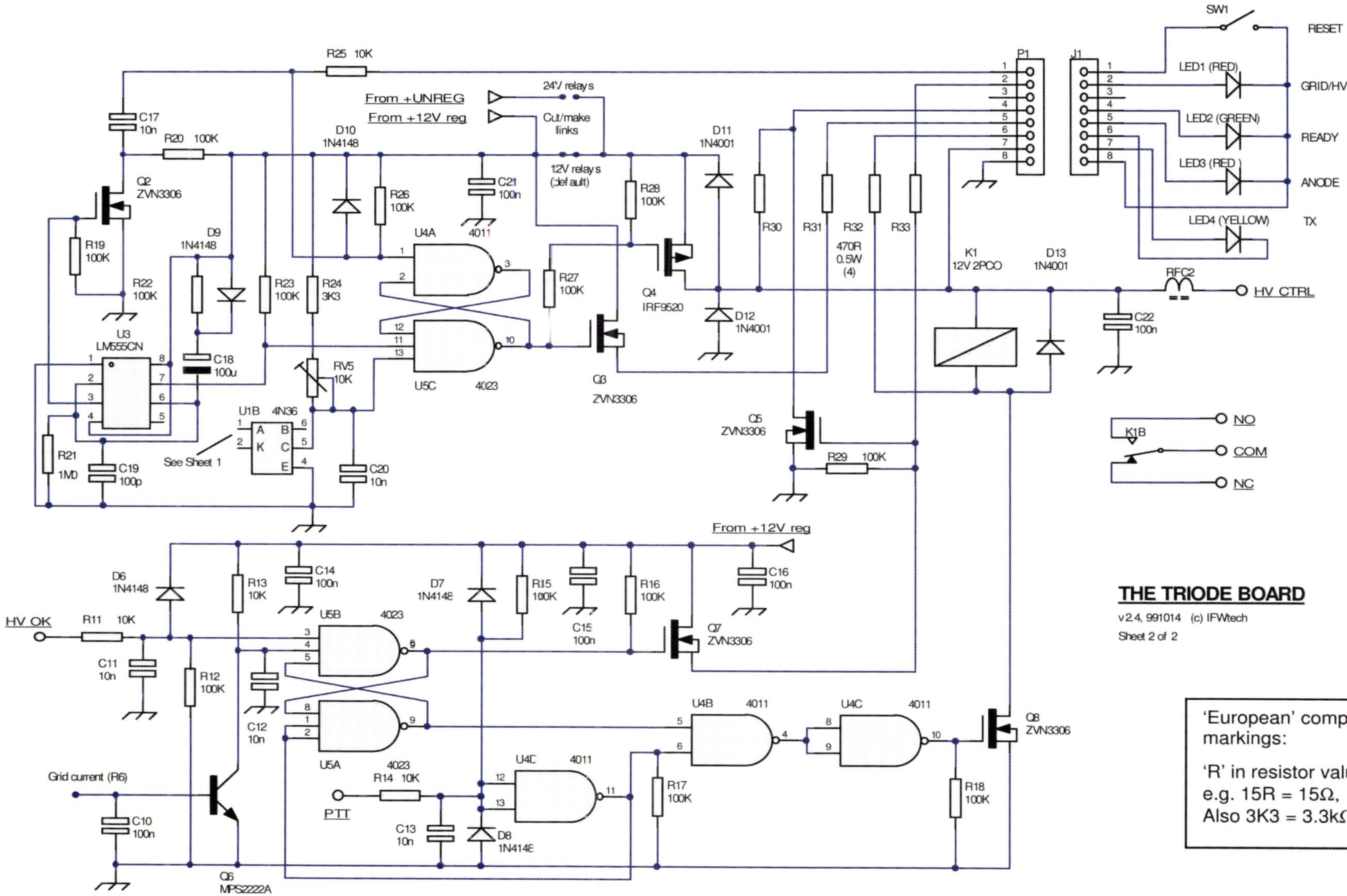

'European' component markings:

'R' in resistor values means Ω, e.g. 15R = 15Ω, 0R33 = 0.33Ω. Also 3K3 = 3.3kΩ etc.

The Triode Board – Interconnections 1

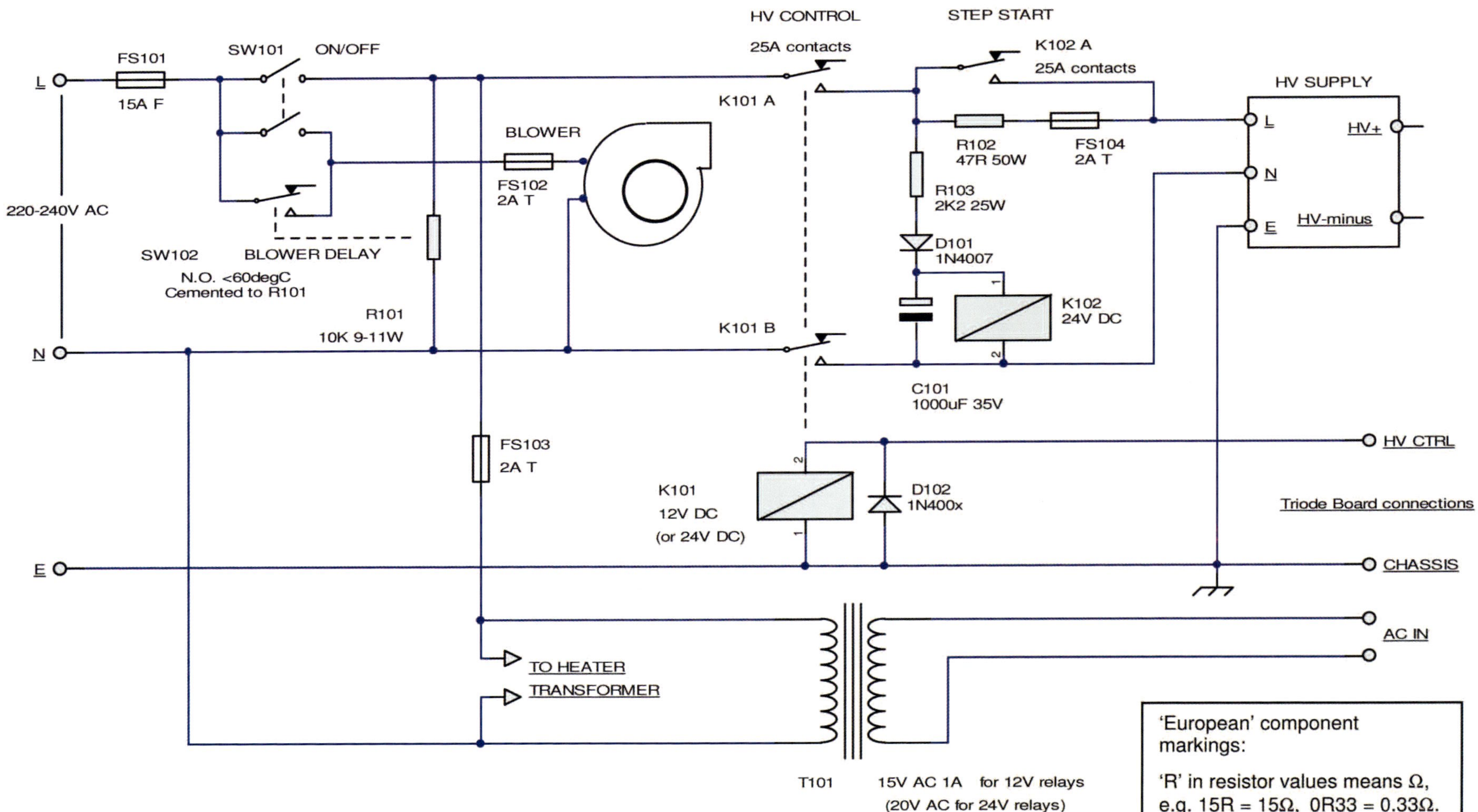

'European' component markings:

'R' in resistor values means Ω, e.g. 15R = 15Ω, 0R33 = 0.33Ω. Also 3K3 = 3.3kΩ etc.

The Triode Board – Interconnections 2

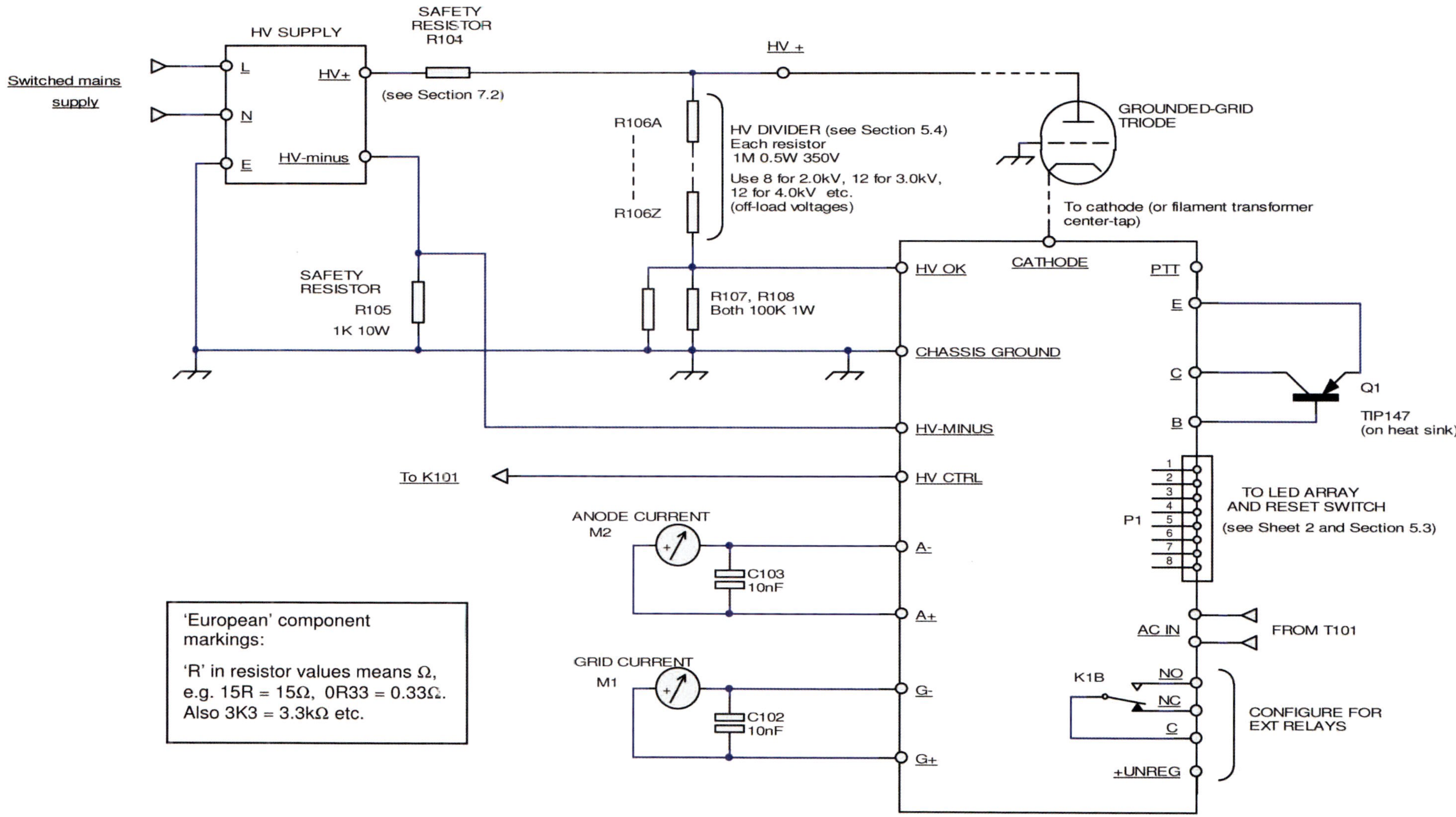

Chapter 13

The Power Supply and Soft Start

The Power Supply
The mains transformer

Iron core

The above iron core transformer is the most commonly used in almost all commercial amplifiers that are made in America or Japan. In larger 1500 watt amplifiers the transformer is usually shipped separately. It is heavy compared with the rest of the amplifier and cabinet. I have seen amplifiers shipped with the transformer still in and dropped as couriers often do from the back of a lorry. The amplifier stops! But the transformer keeps on moving like a missile and destroys the case and any other components in its path.

The transformer comes with large multi pin plugs which connect to the power supply board and heater supplies.

The transformer is often the most expensive part of the amplifier. Getting it replaced can cost into 3 figures plus a heavy shipping cost. This is one reason I moved on to more modern toroidal transformers which are at least 50% lighter and cheaper.

Transformers are very efficient and when there is no load on a transformer they hardly draw any current off the mains supply.

A modern Toroidal Transformer

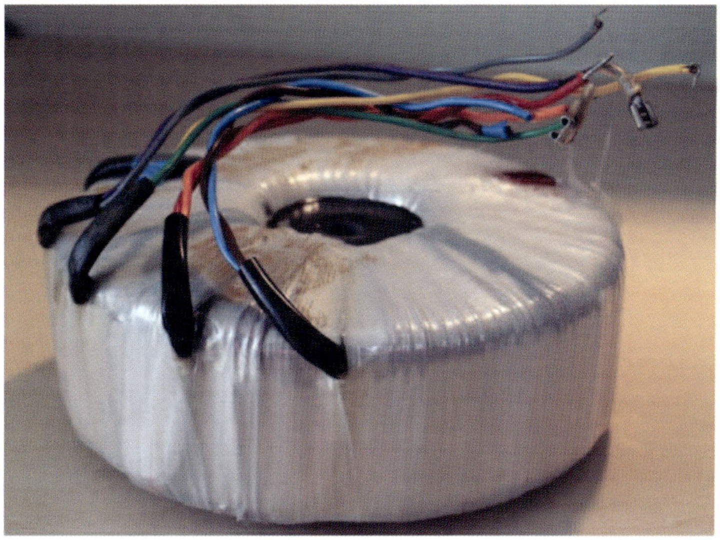

The second choice is a toroidal transformer like the one above. I used my first toroidal transformer in the 2m Explorer and HF Hunter Amplifier around 1985. All Linear Amp UK amplifiers use toroidal transformers. Making a toroidal transformer is a complicated process especially for higher voltages. The primary is wound on to the ferrite core first. Then the secondaries are insulated with a clear plastic film and are wound on the top. There is insulation between each winding. The winding machine is loaded with the correct amount of wire on to a shuttle like a sewing machine. Every turn has to go round with the shuttle through the centre hole sometimes 100s of times to give the required volts per turn.

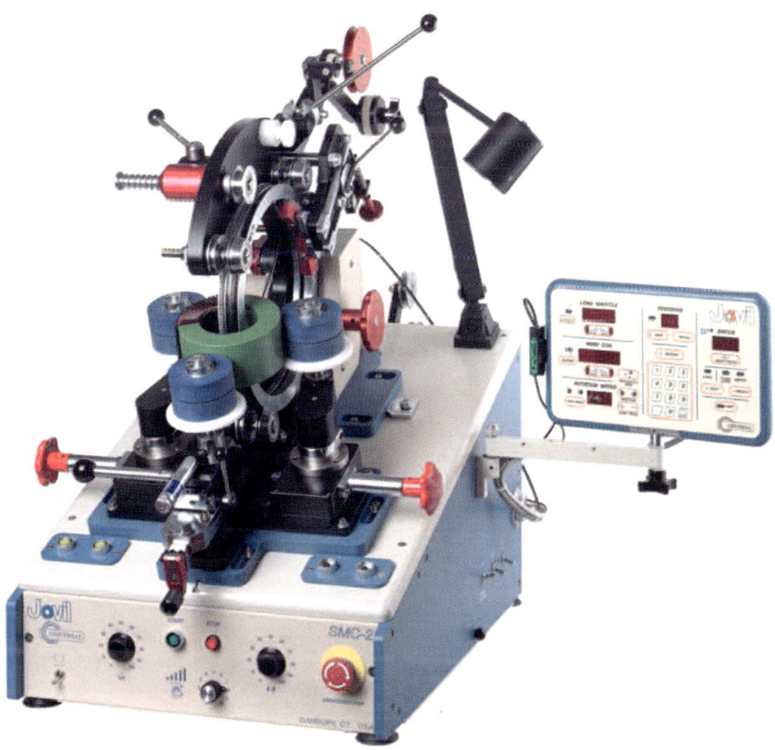

The picture above shows a small toroidal winding machine that would be good for small transformers up to 100 VA

This larger toroidal winding machine is used for larger VA ratings of over 1 KVA

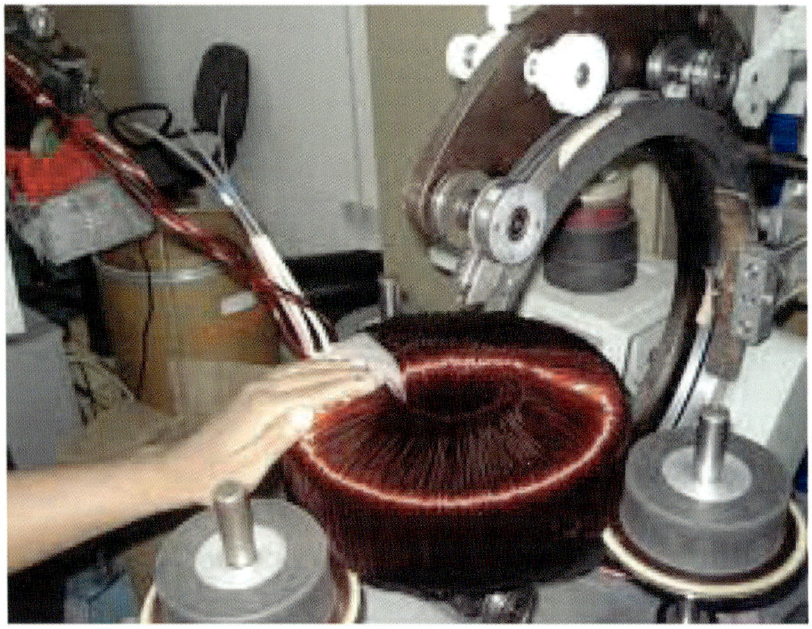

The Tweak factor that you can do to with a Toroidal Transformer

On a toroidal transformer all the windings come to the outer surface which makes it possible to tweak a toroidal transformer to give you the voltage required especially on lower voltage windings. For example if your heater supply gives 5 volts and you require 6.3 volts you can add another 1.5 turns of a similar thickness wire and put it in series. The ratio is usually 1:1 so adding 1.5 turns will normally add around 1.5 volts to give you 6.5 volts AC. The new winding must be in phase. No harm will be done but instead of adding 1.5 volts you will lose 1.5 volts and you will only have 3.5 volts. This can be left if your required voltage is lower than the original. If you need, for example an extra 12V winding use 14 turns and check the voltage when you have done it. You can reduce the winding by 1 turn at a time until you get your desired voltage. Make sure you insulate each turn well especially as it passes through the centre hole and keep them spread out evenly. Fibreglass sleeving or heat shrink tubing is ideal to insulate. Usually toroidal transformers are supplied with a neoprene washer to protect the windings from the outer plate. It is quite absorbent and will cope with small extra windings. If the current for your new winding is only a few amps 1 or 1.5mm enamelled wire will suffice.

Of course using a toroidal transformer you can always strip the secondaries off and individually wind a new low voltage winding on yourself. I would not recommend high voltages unless you are very patient.

Advantages of a Toroidal Transformer

Usually a toroidal transformer is around 33 % of the weight of an iron core transformer and usually at least half the cubic size and cheaper. The cost of a toroidal transformer is usually 50% of the cost of an iron core transformer.

Toroidal transformers don't usually suffer from noise. Older iron core transformers can become loose and give a loud hum. If this happens the transformer can be re-varnished by re-dipping in a bath of hot varnish and left to set overnight. This should bond the windings again and reduce any hum.

I do not know what the maximum size of a toroidal transformer can be. I have bought 5 kVA ones and used them on 3 phase with a 415 volt input. I have noticed that OM with their large 4 and 5 KW amplifiers use toroidal transformers. I am sure for all ham radio applications you should be able to find a manufacturer that will supply one for your requirements. European amplifier manufacturers have been in the lead compared with American and Japanese.

Watts or VA rating

When making an amplifier you will think about the input power in watts.

The transformer will be quoted in VA (volts x amps) there really is not much difference for small transformers. On larger transformers the power factor comes in and has to be catered for. This applies mainly on inductive loads like motors etc. where capacitance is added across the supply to improve the power factor.

For example if you use a GS31b to give you 1200 watts on 2m, your input power will be close to 2000 watts or 2000 VA.

Manufacturers would not use the full 2000 VA input rating.

The transformer would be too large and heavy and impractical. The main danger is overheating of the windings. When the VA rating of the transformer is exceeded a large volt drop occurs due to saturation of the core.

Being Practical

When you have 1000 watts input the required size transformer can be smaller by 50% i.e. the transformer can have rating of 500 to 600 VA. When you are on SSB or CW you spend at least 50% of the time on receive. Also when on SSB or CW the input is less than 50% of a continuous carrier.

This also applies to valves. Take the 3-500Z which has an anode dissipation of 500 watts it can happily be run at 1.5 to 1.8 times its value. Hence the rating of 750 watts output is fine.

Fault finding on a transformer

If you suspect that your transformer is faulty the first thing to do is smell it. A short of any kind will create a hot spot in the transformer and the varnish will put out a strong pungent smell.

The most common fault on a transformer is when a turn gets shorted to the next turn and creates a shorted turn. This will get very hot and smell. A good example is the Weller soldering gun as it uses a single turn wound over the primary. The bit is the narrow bit of the secondary of the one turn coil so gets hot very quickly as the current is high but at a low voltage.

Another fault is when the primary or secondary shorts to another winding. This can be tested with a high voltage insulation tester. You need to test between the two suspected windings. This is usually done with a 400 volt insulation tester and check for any leakage between the windings. Known in the UK as a Mega! The insulation should be at least 1 megohm or more between windings. Anything less makes the winding suspect.

A typical HV voltage doubler power supply

This is a typical voltage doubler power supply rated up to 3200 volts

The capacitors are 220 µF@ 400 VDC giving 27.5 µF of smoothing.
There are 10 x 1N5408 diodes and also 220k discharge resistors across each capacitor.
There are 470k resistors across each diode. This doubler is used in the Ranger 811H and all the GS31 VHF amplifiers.

Capacitors for your Power supply

Voltage working is the working voltage multiplied by the number of capacitors

8 capacitors x 400 volt working has a total of 3200 volts working

The total capacitance is calculated by

$$\frac{1}{C\ total} = \frac{1}{C1} + \frac{1}{C2} + \frac{1}{C3} \text{ etc}$$

Another easy way to work out the capacitance is if you have a 220µF capacitor. Then you add one more in series it comes down to 110 µF. If you then add another 2 making a total of 4 the value halves again and goes down to 55µF and so on.

Electrolytic Smoothing Capacitors

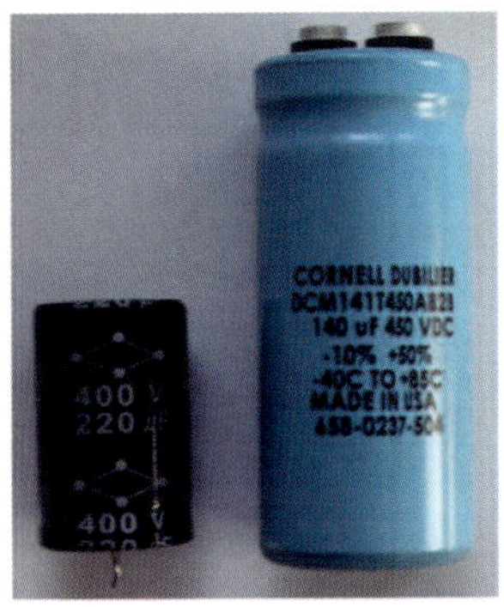

The two electrolytic capacitors above demonstrate the reduction in size over the last 30 years. The smaller one has almost twice the capacitance of the larger one and has a similar voltage working.
If not enough smoothing is used your power supply will suffer volt drop, your audio PEP will be less and distortion will creep in or 50Hz hum will be present on your CW signal. You cannot have too much smoothing! Modern day electrolytics are a fraction of the size and cost of ones 30 years ago. Most amplifiers have around 25 to 35 µF of smoothing on the HT supply. This is achieved by using 6, 8 or 10 x 400 to 500 volt electrolytics in series at between 220 to 400 µF each. A lot of older amplifiers only fitted 100 µF capacitors due to their design. Modern ones are much smaller and it is ok to fit larger capacitance ones that will give you better smoothing and a better PEP output.

Note! When fitting electrolytics in your high voltage power supply, make sure they are fully discharged. It is a must to fit a discharge resistor across the terminals. A good electrolytic can hold its charge for many months. If not discharged properly they can easily give you a nasty shock or throw you across the room, especially if you get across the full smoothing stack.
The lower the value of the resistor, the more current will be taken and the quicker the discharge. You can go as low as 50k but normally 100k to 150k is used at 2 watts rating.
Do work out the ohms law wattage required!

When changing old electrolytics do check that the discharge resistors are all about the same value. If not, a single capacitor can have a much higher voltage across it and will blow straight away.

Capacitors should discharge in 1 to 2 minutes but then short to earth to be doubly sure before touching.

Diodes

Diodes used in a bridge or voltage doubler circuit are very robust and cheap these days. For most amplifiers the 1N5408 is a good choice, they are 1000 volts working @ 3 amps rating. Plenty in each leg is always a good thing. The volt drop across each diode is less than 1 volt, so in a 2000 volt PSU the voltage drop is negligible.

Bridge Rectify or Voltage double

The choice really is yours and at the end of the day it really does not matter.

If you decide to bridge rectify the AC from your transformer it has to be twice the voltage. The advantage of a bridge rectifier is the current drawn is lower and voltage drop of the transformer is less and in theory the regulation is better.

In practice 90% of all commercial amplifiers use a voltage doubler circuit as it keeps the transformer smaller using less turns. A voltage doubler board only needs to give 1 amp at 2500 volts to deliver 2500 watts.

The current from the transformer is hardly a problem, as small as 1mm winding wire is sufficient.

If you choose to bridge rectify, multiply the AC voltage by 1.414 to get the working DC voltage.

For example 1000 volts AC x 1.414 gives a DC working voltage of 1414 VDC

If you choose to voltage double, multiply the AC voltage by 2.84

For example 1000 volts AC x 2.84 gives you a DC working voltage of 2840 DC

An 850 VAC transformer used in the Yaesu FL2100 or Heathkit SB200 uses a doubler circuit to give 2414 VDC, ideal for 572b valves.

Larger amplifiers often switch the voltage of the transformer like the Kenwood TL 922 and Heath SB220 to a CW or SSB position. This only reduces the HT voltage and limits the maximum current input.

It does not alter the bias voltage but will reduce the standing current.

The AC voltage is switched on the front panel between 850 VAC to give 2400 volts DC

1150 VAC will give 3200 VDC. The higher voltage is to make the full 2000 watts input available if required.

Adding soft start to an Amplifier

Adding a soft start circuit is one of the simplest things you can add to an amplifier. It can be fitted internally or external to the amplifier. It should give your valves a longer life. The valve is one of the most expensive parts in your amplifier. Adding a soft start circuit is particularly an advantage to amplifiers that have directly heated filaments like the 3-500Z, 811, and 572b valves.

The reason is that the heater runs at white hot compared with valves which are indirectly heated and run at a dull red. These are more like an electric fire at a dull red colour. This is compared with the bright white of the directly heated valve. For example, if you put 110 volts on a normal light bulb first then after 5 seconds go to 230 volts the filament would hardly ever break, it would be pre-warmed before going to its full power. All the stress is on switch on when the filament is cold.

Valves with the prefix 3CX or 4CX and the Russian triodes and tetrodes would not benefit from soft start to the heater supply. However larger amplifiers would benefit limiting the inrush current to the HT transformer and smoothing capacitors. A soft start will help ease the switch on current through the rectifier diodes and ease the stress on the ON /OFF switch on the front panel or relay.

When I started making the Explorer, I always used a soft start circuit. You get a peak of about 5 amps compared with over 25 amps without it fitted. Amplifiers that would benefit from a soft start fitted would have been the Kenwood TL922, the Heathkit SB220 and all the 572b amplifiers. When you hear the thud from the AC mains transformer, that is the stress that is on switch-on. Internally this applies to the power PCB boards, transformers and the voltage rectifiers.

To fit soft start

You have to cut the live feed going into the amplifier after the on and off switch. Fit a large 50 ohm resistor, the bigger the better up to 200 watts in series which gets shorted out after 3 seconds by the timing relay. I have seen many complicated circuits to do the job but I think the simplest circuit is the best for anybody. It involves a 24 volt DC relay and a diode, a 2k resistor and a 4700 µF capacitor, all shown in the circuit below. But if you wish it can be built into a separate external box with its own on/off switch. You must leave the amplifier mains on/off switch on. Even put some tape over it so you do not switch the amplifier off by accident. Then use only the on/off switch on the soft start unit.

Soft start circuit

Amplifier mains on/off switch

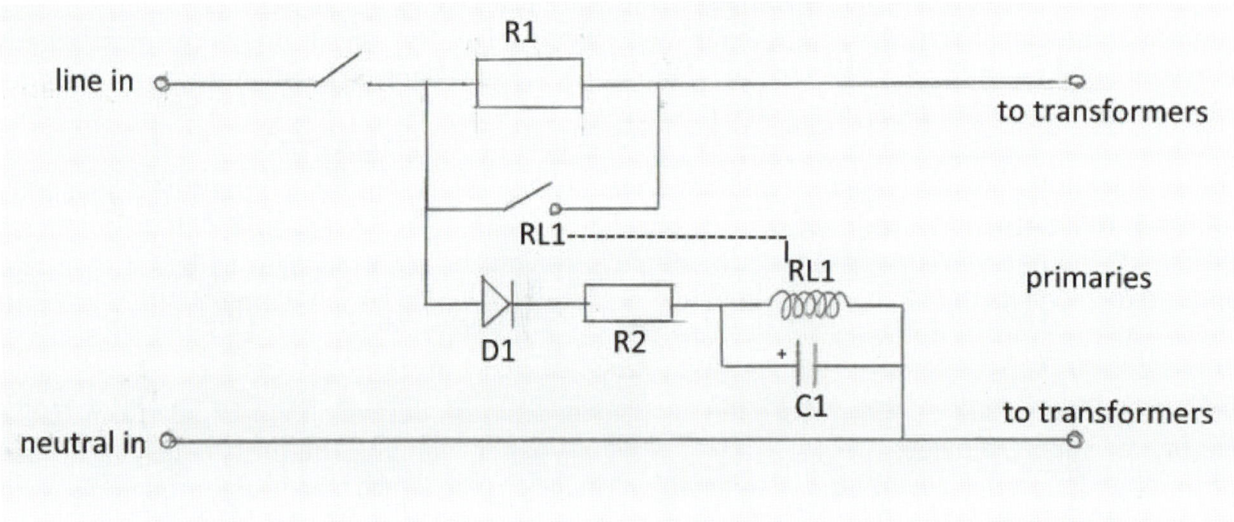

RL1	24 volt single pole DC relay
D1	1N4007
R1	50 ohm 50 to 100 watts
R2	2K2 50 watt
C1	4700 uF

Wiring the circuit is very straight forward and easy to do, the neutral should not be touched except for connecting RL1 and C2 to it. Locate the mains on and off switch and remove the feed going out from it.

Fasten a wire which feeds R1 /RL1/D1 to the switch.

C1 sets the delay and should be around 3 seconds. This can vary as most 24 volt relays have a coil resistance of between 500 and 800 ohms. Normally, the larger the power relay, the lower the resistance of the coil. The relay contacts need to have a handling current of 10 amps or more.

Measures the voltage across C1 or the coil of RL1, it should be close to 24 Volts DC when fully charged, if it is much higher increase R2 to a 3K9.

External soft start unit can be placed between your mains socket and your amplifier.

Chapter 14

Other products made by Linear Amp UK

Balanced ATU

SPC Tuner

Balanced Aerial Tuner 10 -160m

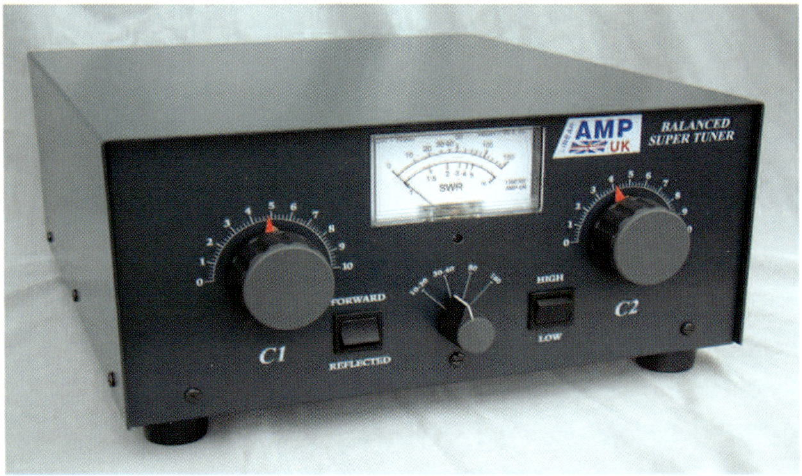

This is a balanced tuner for 10m to 80m, with a tuning range of 30Ω to 1000Ω with low loss. 10m to 80m does not need any switching. If you add 160m it will have to have an extra coil and extra capacitance added. The circuit is two fixed link coupled output circuits. Both are fed with the same input coupling capacitor, a series circuit for 10–20m and a parallel tuned circuit for 30m to 80m or 160m. The balanced tuner, though not intended for unbalanced line, can happily tune a long wire or coax by earthing one side of the link couple. The circuit was originally used by KW Electronics in their EZee match. What I show is a heavier duty tuner that will handle up to 1 kW.

You can feed two separate aerials at the same time. The ATU automatically selects which one is required depending which frequency you are on. For example, you could feed a centre fed LF aerial for 30 – 80m and also feed a full wave loop for better DX operation on the HF bands. The ATU does not need an expensive roller coaster, balun or ceramic switches to operate 10m through to 80m.

Below shows the balanced Aerial Tuning Unit

10-20m coil L1 & L2
50 mm former 60mm long

30m to 160m L3 & L4
70 mm former x 100mm long
(plastic or fibre glass)

Input coupling capacitor

Split stator capacitor, one half 10m to 20m
Second half 30m to 80m or 160m

Balanced Tuner Description

The top left input capacitor is 350pF and is used on all bands. If used on 160m it will sometimes require an extra 250pF fixed capacitor placed across it with a switch or relay S1. This depends if the aerial is of low or high impedance. The current can be large and hence a 250pF door knob capacitor would more suitable. The right split stator capacitor is 2 x 350pF. One half is used for 10 to 20m series circuit and the other half used to resonate the parallel 30 to 160m circuit. This model, though you cannot see it has a double pole relay, both poles are connected in series for tapping the coil for 160m to give a higher voltage working. This will give you a good working voltage and will cope with 1000 watts. On all other bands the relay is closed so the break- down voltage does not matter. A suitable ceramic switch would also work fine.

S3 needs to be switched in for 160m to add an extra 700pF. You will have to play with the value but I find 700pF for C3 means C2b will cover the entire 160m band. I use 1 x 500pF and 2 x 100 pF in parallel. The efficiency is very good compared with most other ATU that use ferrite to give you a balanced output.

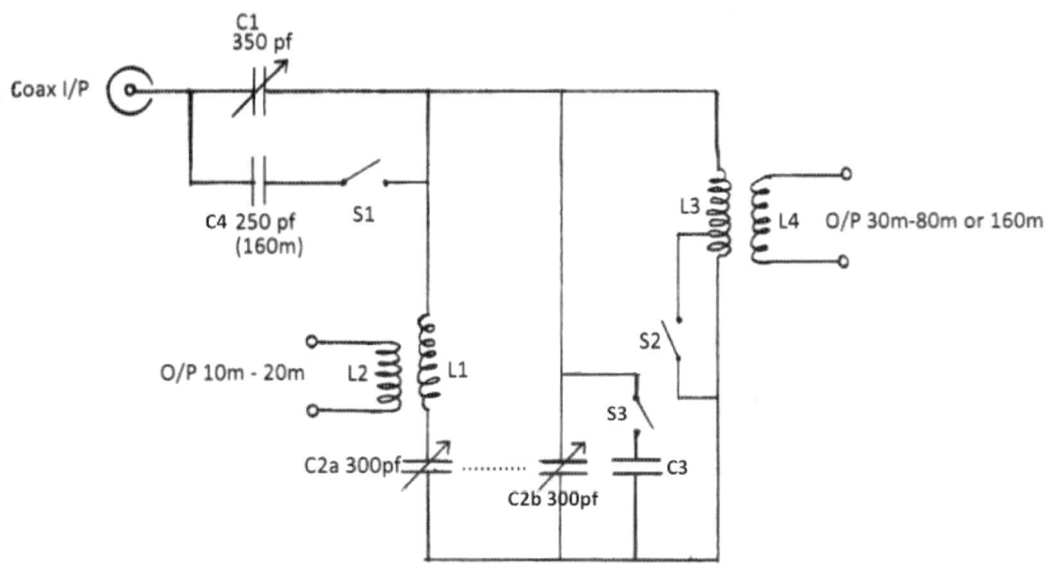

Balanced tuner 10m to 80m or 160m

The SWR power meter can be an external power / SWR meter. I would recommend the Diawa CN101 or CN801. Both are a good quality cross needle meter and rarely go faulty.
By leaving 160m off it makes the ATU easier to make. You will not need S1, S2, S3 C3 and C4

Balanced Tuner case size is:-

325mm wide x 350mm deep x 155mm high
13in wide x 14in deep x 6 in high
Aluminium 1 or 1.5mm thick

Main components of the Balanced Tuner

L1	10m- 20m Primary 4 turns of 2.5mm silvered copper wire, 45mm former wide spaced, (on the inside)
L2	10m-20m Secondary 6 turns of 2.5mm silvered copper wire wound on a 50mm former (placed on the outside)
L3	30m-160m Primary 13 Turns of 2.5mm silvered copper wire on a 70mm former, tapped 0 -- 7--13 turns (wound outside)
L4	30m -160m Secondary 11 Turns of 2.5mm varnished copper wire wound on a 65 mm former (wound inside) (8 turns if you do not need 160m)

C1	350pF wide spaced with ceramic or fibreglass ends, used for coupling on all bands
C2a	350pF half a split stator 10 – 20m series resonator
C2b	350pF half a split stator 30-160m parallel resonator
C3	700pF 1 x 500pF 2 x 100pF in parallel
C4	Optional
S1	used to add extra add extra C on 160m, any type of single pole ceramic switch.
S2	used to shorten the primary coil, 0 –7 turns for 30/40m 0—14 turns for 80m and 160m
S3	Double pole 20 amp Omron relay, both poles connected in series, or a single pole ceramic switch.

You will find the Balanced Tuner will work well on all bands. The efficiency is better than 95% on all bands. The impedance swing can be large. A half wave centre fed will be below 50 ohms to thousands of ohms for open wire feeder. This will determine if you will need to use S2 and the extra 250pF capacitor on 160m. It should be of a high current door knob type capacitor.

SPC or T match Aerial Tuning Unit

Nowadays almost all ATU manufacturers use the SPC-T match configuration. It is very versatile and it will match almost anything. Even if you want to match a grounded tower on 80m or 160m it is perfect choice to feed the gamma wire. The question is whether to use a roller coaster or a ceramic switch. I personally would always use a ceramic switch. Changing bands can be just a few clicks away. Would you dream of putting a roller coaster into an HF amplifier? Never! You have enough scope for impedance change with the two variable capacitors and you do not need a third added if the inductor tap is in the correct position for the band.

The picture above shows my SPC tuner which will tune a long wire, coax or balanced feed line with a ferrite coupler. The input operates at low impedances. The current at higher powers can be large in the inductor and capacitors. I have used a 3mm silver plated copper tube for 10 - 40m and 2mm enamelled copper wire for the extra 80m and 160m part of the coil. The loss is low when measured after the ATU. The two capacitors are 300pF each and the coil is tapped as shown below. This model is fitted with a by-pass circuit which is done with a 20 amp double pole relay. It can be seen in the top left corner, it is wired the same as a Linear Amplifier bypass circuit.

High power ceramic switches are expensive! A good alternative is to use high power switches that are used to switch industrial heaters and motors. They are made of ceramic and will certainly handle the power of a SPC tuner up to 1 kW. These are available from Ohmite Mfg Co., Stokie, Illinois 60076 USA. Look at their web site. The ceramic switch pictured in this ATU is an Ohmite.

Circuit of SPC Tuner

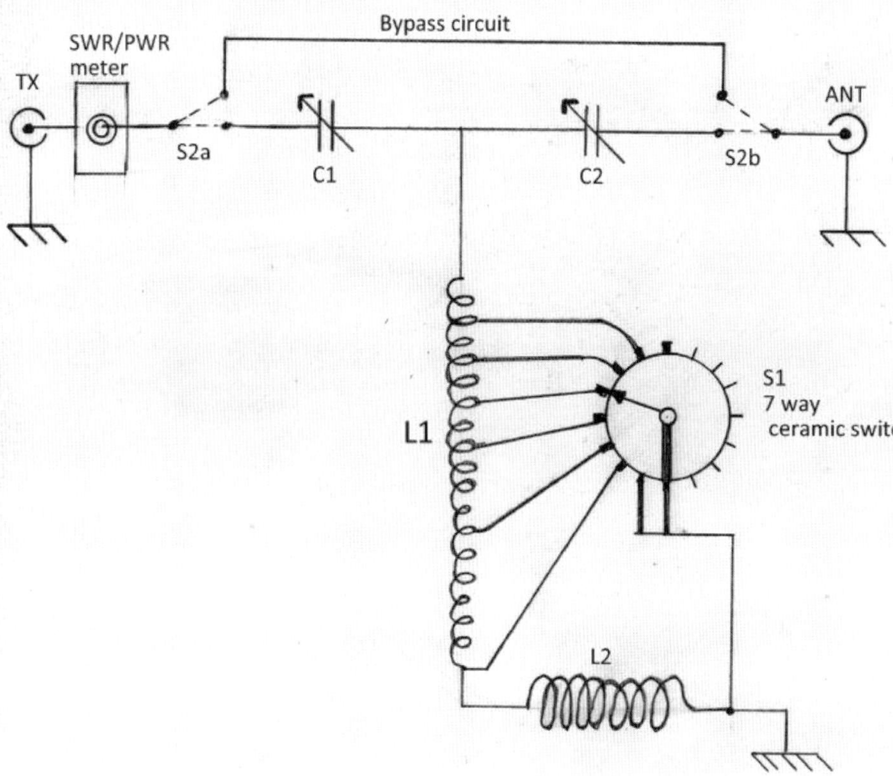

WARNING

Both C1 and C2 are floating above earth and are RF live. A good shaft insulator must be used between the front panel knob and each capacitor.

L1	10m to 80m see following picture and details. 38mm former (1½ inch)
L2	160m inductor 50mm former (2 inch)
C1 and C2	350pF wide spaced variable capacitor
S1	7 way good quality ceramic switch, see text
S2a/b	Optional bypass circuit, double pole ceramic switch.

Alternatively use a double pole 20 amp 12 volt relay and use a small switch on the front panel and feed from your radio 12 volt PSU
SWR/PWR meter can be fitted on the front panel or use an external meter like a Diawa CN101

Alternative

L1, L2 and S1 could be replaced with a single roller coaster. It would need to have approximately 35 turns and be made of 1.5 mm silvered wire, ideally a diameter of 50mm or 2 inch

The main L1 inductor of the SPC ATU

25 Turns total	12t	8t	6t	4 t	2t
Band 80m	40m	30m	20m	17m	10/12m

The left end of L1 is connected to the top of the 160m coil.

The above coil is the main coil and is made of 3mm copper tube. This is available from copper stockists and plumbers merchants. I have silver plated it, but this is mainly cosmetic and could be polished then varnished.

Making a SPC tuner is one of the easiest items you can make and is very versatile for your shack. It will match almost anything. The circuit is used by most almost all ATU manufacturers.

It can be used for matching coax, a long wire or feeding a 4:1 balun. Try feeding your tower as a vertical by feeding the gamma wire from this ATU.

L1 coil – Former size 1½ (38mm)

Method of connecting the tap wire to the 3mm copper tube

Make the tap wire into a hook, place it around the large wire and squeeze together with long nosed pliers. Solder with a large soldering iron or soldering gun. This method can be used up to 6mm tube.

Coil L2

L2 is only used for 160m, it is 20 turns of 1.5 mm varnished copper wire on a 50mm plastic or fibre glass former. The bottom of the coil is fastened to earth and the rotor of the ceramic switch.

The case size is

325 mm wide x 350 mm deep x 155mm high
12 in wide x 14 in deep x 6 in high
Aluminium 1 or 1.5mm thick

How to test a 4:1 balun on all bands

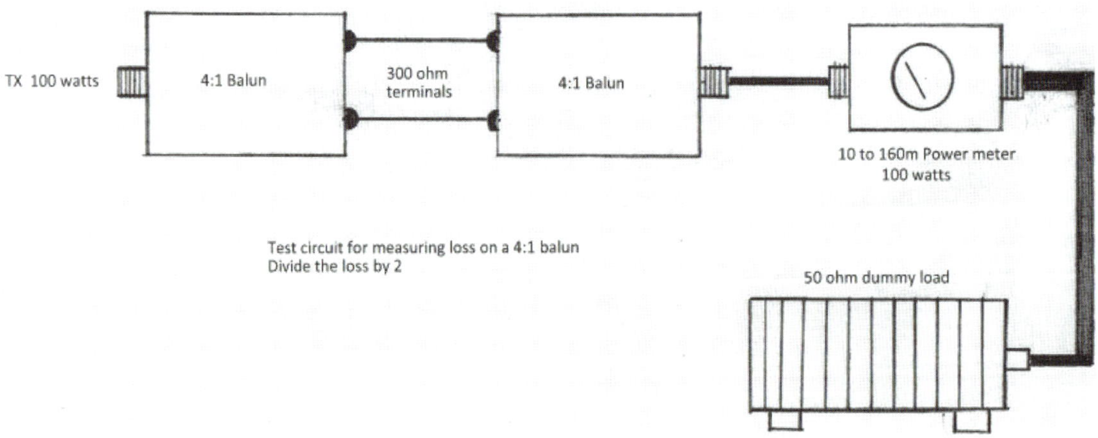

It is very useful to know just how good your 4 to 1 balun is.

Use 2 identical 4:1 baluns and connect the 2 outputs together, i.e. the 300 ohm terminals.

Feed one 50 ohm input into your radio and then measure the output of the other 50 ohm on a power meter.

Check on all the bands 10 to 160m! Half the loss shown loss on the meter and you will know how good your balun is on each band.

This book is available
at £30.00 plus £5.00 post and packing
(cheque, cash or bank transfer)

from :-

Peter Rodmell
2 Meadow Way
Walkington
Beverley
East Yorkshire
HU17 8SD

e mail g3zrs@hotmail.co.uk
phone 01482 862323

also available on e bay (pay by PayPal) under Linear Amplifiers

also stocked by The RSGB, DARC, and ARRL